D0839767

EARTH EMOTIONS

EARTH EMOTIONS

New Words for a New World

Glenn A. Albrecht

Cornell University Press
Ithaca and London

First published 2019 by Cornell University Press

Printed in the United States of America

Library of Congress Cataloging-in-Publication Data

Names: Albrecht, Glenn, author.
Title: Earth emotions : new words for a new world / Glenn A. Albrecht.
Description: Ithaca : Cornell University Press, 2019. | Includes bibliographical references and index.
Identifiers: LCCN 2018045107 (print) | LCCN 2018055649 (ebook) | ISBN 9781501715242 (pdf) | ISBN 9781501715235 (epub/mobi) | ISBN 9781501715228 | ISBN 9781501715228 (pbk. ; alk. paper)
Subjects: LCSH: Human ecology—Philosophy. | Environmentalism—Philosophy. | Nature—Effect of human beings on. | Emotions—Health aspects.
Classification: LCC GF21 (ebook) | LCC GF21. A44 2019 (print) | DDC 304.2—dc23
LC record available at https://lccn.loc.gov/2018045107

For Lilly, Teddy, Lyra, and Generation Symbiocene

Contents

Preface

My lifetime coincides with the massive transformational forces of the Anthropocene, the period of human dominance over all biophysical processes on the planet, including the big one, a hotter and more chaotic global climate. The effects of social and biophysical change are now so large and rapid that it is difficult to comprehend them all.

As I have matured into this life I feel capable of reflecting on what it means to live in the Anthropocene. In this book, I have concentrated on one major theme: the particular human emotional responses we have to the scale and pace of ecological and environmental change. I call these responses "Earth emotions."

Writing a book about Earth emotions was bound to be an emotional experience, and I have constructed my own "sumbiography," or an account of the sum total of events and experiences that have coalesced to produce my particular, and perhaps peculiar, take on nature and life. Through this telling, I hope to help readers understand and respond to the huge turmoil we are in, as a species, at this moment of Earth's history. I also want to help create a way out of it.

I have also drawn on older cultural traditions to understand what our relationship to the Earth is about. Australian Aboriginal people have been in the "Great South Land" for up to eighty thousand years. They have a powerful story of longevity and coexistence with other humans and nonhumans to share with the rest of the world. Many Aboriginal elders have a double experience of turmoil. First their traditional culture was dismantled by colonial forces; and now, their emergent hybrid culture is assaulted by climate chaos. Their experience of loss and persistence is instructive to all humans.

As the threat to humanity of a powerful climatic *global* force of our own making has never before existed, much of what has been written in the past about nature and life is irrelevant to the future we now face. Also, since the essence of the Anthropocene seems to be self-destructive, I want to exit the era as soon as possible. *Earth Emotions* will take us directly to the core of our problems and use the knowledge gained by the latest science to guide us rapidly into a better future.

Based on the biosciences that underpin discoveries about the central role of symbiosis (living together) in the lifeworld, I have created "the Symbiocene," an alternative "good" future place with distinctive positive Earth emotions. Along the way, I create many new ideas, words, and concepts that I think will challenge the representatives of the Anthropocene and usher in the Symbiocene.

The first of these concepts is "solastalgia," or the lived experience of distressing, negative environmental change. I tell the story of the "discovery" of solastalgia and how it has become a key concept in the domain of human responses to negative and distressing environmental change. Solastalgia is as relevant to individuals' loss of an endemic sense of place due to the negative impacts of global warming as it is to cities and their urban complexes as they are transformed by the forces of development.

After creating the concept of solastalgia, I came to the realization that this negative emotional and psychological experience sat within a wider range of what I now call "psychoterratic" (psyche-earth) emotional concepts. I have set about describing these negative Earth emotions and juxtaposing them against positive ones. I call the result of this comparative analysis the "psychoterratic typology," and it is work in progress as the novel effects of grand-scale environmental damage to places, hearts, and psyches become evident. Conversely, as we lose beautiful and loved places on Earth, we begin to put into words those positive emotional states that were once delivered for free and without the need for terminology. Positive and negative Earth emotions are in lockstep, as you cannot have one without the other. I have devoted two chapters in this book to the explication of these dialectically opposed elements of the psychoterratic typology. Included in these chapters are the ideas of other writers and thinkers who have contributed to the typology. Some, like Aldo Leopold, are famous, while another, Elyne Mitchell, an Australian writer and a contemporary of Leopold's, is almost unknown as an environmental thinker even within Australia.

The revolution in scientific and transdisciplinary thinking, based on the foundation of life we call symbiosis, is also a stimulus for the creation of a new secular form of spirituality that accompanies the science, technology, ethics, and culture of the Symbiocene. I have named this new secular spirituality "the ghedeist," and it, too, is vital for the development of a new human relationship with the rest of life. I am of the view that many of my readers will feel, like me, that the old sources of spiritual insight are not going to make the grade in the twenty-first century. If you crave a new secular spirituality that reunites humanity with the Earth from which we emerged as a species, then I have something to offer you.

The book concludes with an examination of what humans must do to enter the Symbiocene. Although a transdisciplinary philosopher by inclination, I have attempted to give "Generation Symbiocene" a relentless, optimistic, and practical view of their future. Such extreme optimism is needed to counter the ruthless pessimism that is emerging from the critics of the Anthropocene and the collective wisdom of scientists who keep warning us about the apocalyptic scenario that is unfolding due to ecosystem destruction, toxic pollution, and climate warming. Even baby boomers are given a satisfactory way of exiting this life, as they can play a huge role in enabling their own children and grandchildren to successfully enter the Symbiocene.

I do not make light of the situation we now find ourselves in. I have declared that there is now an open emotional war occurring between the forces of creation and the forces of destruction on this Earth. World War Three will be an emotional war that must end in victory for the forces of creation. Yet those who command the forces of destruction will not easily cede their control of the Earth. The future could turn out to be very ugly for humanity—but then again, it could be brilliant.

Earth Emotions is my contribution to an optimistic future, one that challenges everything that is going wrong on the Earth right now, and it offers solutions. I trust that feelings of solastalgia will become a distant memory of a past we choose to negate. My hope is that, by the end of this book, you will feel totally at home with the new concepts I have created. For one, you will be "in" the Symbiocene: you will feel that you are already part of it, and that it is a part of you.

Acknowledgments

I thank my editor, Kitty Liu, and the editorial team at Cornell University Press for helping me bring this book to fruition.

Having worked as a university scholar in the past, I also thank my past employers, the University of Newcastle and Murdoch University, for allowing me the freedom to think and work in ways that defied conventional discipline boundaries. Helping me in that task have been my friends and transdisciplinary colleagues Nick Higginbotham and Linda Connor and their willingness to tolerate my unconventional ways of thinking. I am also an honorary associate in the School of Geosciences, Faculty of Science, at the University of Sydney. I thank them for their support. A special thanks goes to Jules Pretty, who has supported and advised me on getting my ideas into print.

The people of the Hunter Valley and Gloucester in New South Wales have also been very generous in the time they have given to me in my quest to understand Earth emotions. I thank them for allowing me to share their sense of place and how it has been impacted by large-scale environmental change. I am also in debt to the many hundreds of people worldwide who have been moved by my work in the past and have corresponded with me about their own emotional confrontation with a negatively changing world. Their insights have helped me build the case for the emergent field of the psychoterratic.

I thank my family and in particular my wife, Jill, for tolerating my eccentricities and time spent alone thinking and writing. Jill has helped me with the manuscript genesis and correction of my errors and poor writing. As always, despite her best efforts, for all the remaining errors and omissions in *Earth Emotions*, I take full responsibility.

Finally, it is only with the wisdom of hindsight given to me by virtue of my age (sixty-five at the time of writing) that I have felt capable of tackling a topic as broad and complex as human emotional relationships to the Earth. In engaging in this task, I have been influenced by many great thinkers of the past, and it is my hope that this work reflects their cumulative impact on my thinking.

I thank the owl of Minerva, the goddess of wisdom, for giving me the insight I needed to see this book to completion. As G. W. F. Hegel observed, she "spreads her wings only with the falling of dusk." However, I also thank the original people of Australia and their totemic bird, the Kookaburra of enlightenment. Aboriginal stories say it is a bird in Australia that laughs out loud only with the breaking of dawn.

EARTH EMOTIONS

Introduction

Our origins as a species lie in the universe, which is at once both orderly and chaotic. The more we understand the nature of our universe and cosmos, the more we see that life on planet Earth is the outcome of mighty forces that both build and destroy over billions of years. The universe is a place of restless and endless motion.

The word "emotion" has its origins in the Latin *movēre*, "to move," and *ēmovēre*, "to agitate, disturb." I contend, therefore, that this makes the universe an "emotional" place. Emotions are defined as "that which moves us" or affects us. Our universe is shaped by, and shapes with, powerful forces, and is the prime mover in both creating and destroying the conditions for life. In the most real sense, the universe is the source of all emotional forces.

The emotional forces that characterize the universe sit within the poles of what I call *terraphthora* (Earth destroyer) and *terranascia* (Earth creator).[1] The universe displays its terraphthoric characteristics in the violence of destructive forces. There are collisions between galaxies on the grand scale, rogue comets and meteorites that smash into little planets like the Earth, big asteroids that plaster pockmarks into the face of the Moon. On planets like Earth there are the forces of volcanism and tectonics that blow places apart and cause the land to quake (as if in fright). There are also creative, terranascient forces that build, as spiral galaxies become more complex over time, dust and ice clouds coalesce into planets, chemicals bond to form the patterns of life, and complex organic forms, such as bacteria, fungi, trees, whales, and humans, emerge. We need to pay attention to both the creative and destructive forces in our universe.

The physics and chemistry of our universe are still being researched, so what we understand now is subject to further, perhaps revolutionary, change. Even the venerable big bang theory of the origins of our universe has been contested in the last decade. No single big bang perhaps, no beginning, no end, and the idea of coexisting multiverses is now being openly discussed. Perhaps how our universe was created does not matter? What does matter is that, irrespective of the origins of our universe, humans sit within it as a reflection of its past structures and processes. We are products of a larger system that had its own existence long before ours.

Within this cosmic drama, our ancestors appeared on Earth about 2.6 million years ago, long after the formation of the solar system and the Earth, 4 to 5 billion years ago. For only the last three hundred thousand years, *Homo sapiens*, the thinking or wise mammal, has taken on the task of trying to understand the origins and persistence of all forms of life. That we, as humans, can retrospectively view this cosmic evolution is itself almost miraculous. It would seem that the emergence and persistence of life is a very rare occurrence, perhaps even a unique event, in the long history of the universe. We are a tiny speck of life fighting against the entropic forces of destruction and decay, by building organic order in cooperation with other life forms, against a background of what appears to be thermodynamic decline.

Human Endurance and the Dreaming

In the last ten thousand years, within a geological period known as the Holocene, humans have had a dream run of conditions for their cultural, agricultural, and technological evolution. The combination of a favorable and stable climate, and the opportunity for the terranascient human emotions to prevail over many thousands of years, meant that human social evolution could be consolidated and refined. Phenology, or the patterns, cycles, and rhythms of nature, has been very kind to us, enabling, in particular, the agricultural revolution that delivered permanent settlements and ultimately the rise of cities. Only in the last few hundred years, however, has the possibility existed that humans could wipe out themselves, and most other life forms with them, under global industrial development pressures, by nuclear annihilation and now by climate warming. We need an explanation of how this state of affairs came about.

The interaction between terraphthoric and terranascient emotions has always been a part of human history, and every human culture has its own version of this ancient struggle and ways of expressing it. The emotional makeup of humans is largely locked into the struggle, and we can appreciate that common emotions are connected to both destructive terraphthoric forces and to those that might be associated with creative terranascient forces. In order to defend one's own life, and those of close kin, emotions such as anger, hostility, fear, envy, jealousy, and contempt would have to be primordially motivational. In order for humans to be secure within family and kinship groups, emotions such as love, care, empathy, admiration, and happiness would have to find an outlet.

While both sets of emotions exist, and are commonly expressed, humans (so far) have built their success as a species largely on the terranascient emotions. The nurturing emotions must prevail over the destructive emotions in the long run; otherwise, humans would have wiped themselves off the face of the Earth ages ago. Human infants have an extended period of gestation and maturation and must be protected and nurtured by their parents and kin over a decade or more. The terranascient emotions are expressed as an ethic of love, care, and responsibility, and they are directed at the protection of life—all life.

In the Aboriginal people of Australia, one of the oldest, documented continuous cultures on the planet, we can see the interplay between both sets of emotions demonstrated in their cultural foundations. With a continuous presence of up to eighty thousand years, their longevity and resilience as inhabitants of continental Australia and its islands suggest that, as Earth creators, they were, until European colonization, an instructive example of human success.[2] Their ancient Dreaming stories represent a total conception of nature from the largest scale to the smallest within the context of deep time. To meet human needs, their cosmology is both emotional and ethical, one based on a spiritual dualism of good and evil, good emotions and bad emotions.

The knowledge of astronomy and ecology in all Aboriginal peoples throughout Australia indicates a profound understanding of the phenology of life. Human life sits within larger patterns and rhythms. This emotional choreography was also passed on through generations of Aboriginal people in the oral traditions of the Dreaming stories. The Dreaming, as explained by anthropologists such as William Stanner, was an account of the creation of the Earth and the place of all life within it. Stanner saw three major elements within the matrix of the Dreaming: the great marvels of the biophysical world, the common ancestry of all life and species, and the rules of social life.[3] The first element concerns "the great marvels—how all the fire and water in the world were stolen and recaptured; . . . how the hills, rivers, and waterholes were made; how the sun, moon, and stars were set upon their courses."[4]

Deborah Bird Rose also writes about the Dreaming and, while noting the cultural, linguistic, and academic nuances connected to this term, summarizes the second element by suggesting that it provides a profound awareness of connectedness in all life and living systems and, as a consequence, deep human kinship with all living things.[5] The third element, how humans should live and how their institutions should operate, connects culture to nature in a seamless totality. Earth emotions are what make us human-in-nature.

The emotions of making a living are illustrated in every Dreaming story told by Aboriginal people all over Australia. For example, the Aboriginal people had a profound understanding of the astronomy of the Southern Hemisphere, where all was interpreted as interconnected and mutually interactive. The patterns and images of the stars were used to illustrate aspects of human culture, while the positions of certain stars and their constellations were used to predict seasonal variation and the changes in ecosystems that affect human sustenance. Each bioregion and its human culture understood the seasonal relationship as a survival almanac that tells what is in flower, what is available to be eaten, and where to go next for food.

Lyra, the Almanac, and Human Emotions

The story of the Lyra constellation for the Boorong Aboriginal people around Lake Tyrell in northwestern Victoria makes for an insightful connection between the location of stars in the night sky and the ability to predict and find a valued food source.[6] One of the stars in the Boorong night sky was called Neilloan. It is named after a ground-dwelling, fowl-sized, mound-building bird, the Malleefowl (*Leipoa ocellata*); its Aboriginal name is *Lowan* or *Loan*.[7] The star is part of a constellation that contemporary astronomers call "Lyra." For the Boorong people, its collection of stars have the general shape of a bird with Malleefowl-like characteristics, including a star in a position that gives the appearance of a large foot or leg issuing from the body of the bird.

The Malleefowl buries its eggs in a large mound of sand and decaying matter on the ground. It creates this mound by scratching the earth with its strong legs and by raking leaves and earth into a big pile. Birds that reproduce in this manner are called "megapodes," or mound builders, with the word literally meaning "large foot." The decomposition of the organic matter in the mound generates warmth that enables the Malleefowl eggs to be incubated to the point of hatching. The hatched Malleefowl chicks must then dig their way to the surface of their mound and immediately live an independent life in the harsh, arid environment. John Morieson explains the almanac connections between the bird and the Lyra constellation: "Lyra appears in the southern hemisphere only between March and October, coinciding with the mound building period of the Malleefowl. This is the first of a series of remarkable parallels between the bird in the sky and the bird on the ground."[8]

The next parallel is the way the behavior of the bird is linked to an annual event in the Lyra constellation. The Lyrids is the name given to meteor showers in this constellation that can be seen in April, and "they remind us of the bits of sand, twigs and other matter flying through the air as the Malleefowl kicks material on or away from the mound."[9] The constellation not only looks like the bird, it behaves like one. Finally, just as Neilloan fades in the southern sky in October, the Loan's eggs will be ready to harvest.

The story of the Malleefowl in the night sky indicates an acute knowledge of the timing of events between the cosmos and the Earth. It also connects human emotions with the cosmos and the bird. Human care and empathy intersect with the Malleefowl's phenology and fecundity, which must be respected, or else it would be lost to the ecosystem. For this reason, some people in the Boorong clan would have had the Malleefowl as their totem. The special empathy with the bird meant not only that they could not eat it, but also that a major role in their life would be to look after its habitat. Telling the story of the Malleefowl in the sky is an account of how to live and how to relate to other beings. There is an emotional astronomy that, once understood, gives to people an intimate empathy with fellow creatures that in turn give them sustenance.

Emotional Turmoil

The Dreaming stories often have both peaceful and violent aspects. Sometimes there is a resolution of these opposing forces that allows a successful space for human life to flourish. Many stories simply reflect the emotional turmoil that often occurs in human communities yet is contained within a cosmic background. The explanation of the suddenness by which human life can be taken has direct connections to human emotions, as is the case with the emotions of the creation of life. Dreaming stories from the people of Melville Island and Northern Arnhem Land capture the essence of these primordial forces and their impact on humans. Of thunderstorms, the Melville Islanders say there is "a woman, Bumerali, who strikes the ground with her stone axes mounted on long flexible handles. These are the lightning flashes which destroy the trees and sometimes the aborigines." And in Arnhem Land it is told that "the thunder-man, Jambuwul, travels from place-to-place on the large cumulus clouds of the wet season, shedding life-giving rain on the earth beneath. These thunder-clouds are also the home of tiny spirit children, the yurtus, who travel on the rain-drops to descend to the earth to find a human mother."[10]

Other stories relate states of human affairs to the formation of the cosmos. The Melville Island people tell that, in the ancient past, the men of the Maludaianini tribe had a problem caused by the sexual relationships between men and women. The issue revolved around

> the men . . . always sneaking into the jungle with the women of other men, even though they had wives of their own. This behaviour caused a great deal of jealousy and bickering that finally developed into a fight in which some of the men were killed. After this the Maludaianini people went into the sky, the men becoming the Milky Way, the women some of the nearby stars.[11]

In these stories, human emotions are embedded in the structure of the cosmos, and they assist humans, at any time in history, to see that the emotions generating conflict, such as jealousy and anger, can find resolution and that humans are not perfect. There is a recognition within this ancient culture that human nature is complex with a dynamic mix of violent and peaceful attributes. They did not view themselves as fulfilling the requirements of the trope of the "noble savage."

In addition to seeing relationships between human affairs and the order in the cosmos, as Rose pointed out, Aboriginal people wove their own lives into the lives of fellow creatures and the total landscape. They saw human life as originating in common ancestors that consisted of human and non-human life in a shared destiny. I am sure that similar reciprocal relationships between the cosmos, the Earth, human and nonhuman life, and the physical environment exist within the cultures of indigenous people all over the world.

Therefore, for traditional peoples, order and disorder in the cosmos became a basis for understanding and explaining order and disorder in human affairs. Emotional states such as jealousy and anger find their correlates in the capriciousness and violence in nature and an acknowledgment that they will never be completely eliminated or controlled in all circumstances. Conversely, emotions such as caring and nurturing can also be found in an orderly universe, and in the ways other Earth-bound living beings share the instincts, patterns, and urges that propel life, sex, birth, hunger, safety, and death.

In the Australian context, the forces disturbing climate and the environment reached a turning point at the last maximum ice age. This was in the upper Pleistocene epoch about twenty thousand years ago. As that ice age ended, the land ice melted, and over the next thirteen thousand years, the sea level began to rise from a low point of about 120 meters below the present level. By about seven thousand years ago the sea had reached its present level. Although confronted by a major shift in the shoreline, Aboriginal people had plenty of time to adapt to these changes. The climate and sea level stabilized, and in the Holocene it is generally agreed that, worldwide, humans were able to flourish as a species because of the relative stability of their home environments. This stability was thought to be in place until the planetary-scale disturbance of the Anthropocene after the 1950s.[12]

Aboriginal culture contains an ancient memory of the gradual coastal flooding in the past, and it served as a constant reminder that stability should not be taken for granted. It also offered proof that humans can come out of periods of adversity and be optimistic about the future.

Aboriginal people passed down their emotional narratives about life from one generation to the next. Although technically a nonliterate culture, information was passed on through cultural practices such as Dreaming stories, music, dance, song, and art. Sandstone cave galleries, for example, acted like family and natural-history photograph albums containing the images of the ages, including the ocher mouth-spatter handprints of known ancestors long gone. Often the handprint would be their own, preserved from childhood.

As they occupied the whole of the continent, Aboriginal Australians lived within what we now call bioregions. Within these bioregions they developed a detailed knowledge of their place, or "country." The Dreaming stories of different clans are specific to place and gave them a unique emotional compass for living, one that could be used for thousands of years within a relatively stable home. It is the longevity of this culture, one that incorporated the forces of nature into a spiritual, ecological, and social coherent whole, that I will come back to in subsequent chapters. Industrial human culture, the Anthropocene, has been around for only three hundred years and, by contrast, Australian Aborigines have passed the endurance test for tens of thousands of years.

The arrival of colonists from Europe to Australia in 1788 changed everything for Aboriginal people. Every single aspect of their culture was forcefully overruled by the colonial power. Their spirituality in the form of the Dreaming, their ability to move freely over their own "country," and their ability to control their own destinies were systematically taken away from them. Their emotional foundation was shattered, and it remains an ongoing humanitarian disaster.

I will say more about the emotional desolation of Aboriginal society in later chapters, but I argue that, now, all contemporary humans are in a position similar to that of the Aboriginal people in Australia after 1788. The Anthropocene has arrived as a kind of colonizing force pushing against all previous forms of human culture. A single, homogeneous value system is now influencing our emotional expression and identity in new places, spaces, and contexts. Our ancient human emotional bearings with respect to the Earth, as home, have become universally confounded. Now, it seems, there is only one way of being human.

Losing Our Emotional Bearings

In order to think briefly about the impact of this emotional disturbance, we can speculate that, prior to the arrival of the "sweet spot" of the Holocene, life for most humans was defined by the vicissitudes of scarcity and a large degree of uncertainty.[13] The terraphthoric emotions would have been in constant use, as survival in a time of unpredictable change, with constant threats from other species and competition for resources, must have been dependent on humans having a set of emotional and physical attributes that enabled them to endure. Both men and women, individually and collectively, could give free rein to their terraphthoric emotions when the need to defend home territories from rivals, both human and nonhuman, had to be confronted. Men especially, but not exclusively, in the context of a hunting and gathering economy, also required the emotional attributes needed to hunt and kill animals that, in many instances, could kill or injure them. They also had the attributes, and the occasional need, to kill other human beings.

In addition, detailed knowledge of the home environment was vital to ongoing relationships to places and the elements of sustenance. The terranascient, cooperative emotions were needed to sustain family and clan and must have also been foundational for survival. Obviously, female terranascient knowledge and praxis (actions) were foundational and fundamental for human life to endure.

At some point, the destructive and creative emotional forces had to be brought under control with, all things being equal, the peaceful and nurturing forces prevailing over destructive and violent ones. As with the Aboriginal people of Australia, peoples all over the planet had to achieve, at a minimum, a workable relationship between these two opposing sets of emotions in order for there to be no net loss of people in a resource-constrained environment. Nonlethal forms of justice and retribution were also carefully worked out and applied. Cooperative relationships with neighboring clans had to be cultivated and enforced, since practices such as fire-stick farming meant that fire management and control were in the common interest of all.[14] The natural resources, such as animals and plants, also had to be actively managed by human emotional bonds, in the form of protective personal totems, with those living things that gave them sustenance. Specific people within a clan had emotional bonds that prohibited the eating of their particular totem animal and entailed obligations that would protect the "increase sites" (habitat) of that species. By spreading such totemic identification around a large number of species, the emotional symbiosis acted to conserve the total environment.

While there are many modern contexts where the scarcity-based terraphthoric emotions are still expressed, such as mining, war, and sport, for the most part they have, until very recently, been masked by the achievement of material abundance. The last sixty-five years (my lifetime) have been relatively peaceful, for unlike the experience of global war in my grandparents' and my parents' lifetime, there has been no World War Three. For people in advanced industrial and technological societies, material abundance has enabled a period of global peace and population increase that seemed like it would never end.[15]

That has all changed in the second decade of the twentieth-first century. The tensions between terraphthoric and terranascient forces are once more being played out openly, in a period of rapid cultural and biophysical turmoil. Even our political landscape is now one where often terraphthoric human emotions largely rule the realm of policy and ideas. In many Western countries, predominately ruled by free-market forces, there is even a form of hostility toward the environment being orchestrated by political elites. Fracking, drilling for oil in the Arctic, clear-felling the Amazon for cattle ranching, and the loss of tropical forests for palm oil all indicate something beyond indifference to the environment we live in.

The fabulous phenology of the Holocene has now all but gone. We are entering a period of extreme uncertainty, with massive famines, international terrorism, horrendous nuclear accidents, threats of nuclear war, political extremism, and climate catastrophes all adding to our current anxieties and future dread. Already there is solastalgia for that which is being lost and global grief for that which has been extirpated.

I fear that the terraphthoric emotions are once again rising to the top of our ancient emotional tree. Old institutional sources of emotional support for humans are also failing. For instance, the Roman Catholic Church, despite the efforts of champions of the terranascient such as Pope Francis, is not able to prevent global-scale development that is tearing apart the old emotional order.[16]

New places, such as cities and urban complexes, where more than half the world's population now live, are also locations that deliver complex culturally and technologically defined connections to place. However, along with the loss of the night sky to bright, artificial lights, people lose their emotional compass for enduring within the confines of the living forms and processes of the Earth, the cosmos, and the multiverse. Humans lose their emotional astronomy, emotional ecology, and, with them, their spiritual ecology.

Many have gained material affluence, but at the expense of emotional and psychic security. We are emotionally lost, and those who are in control are using terraphthoric emotions to dominate and control an Earth destiny that reflects their self-interest. Some of them even have visions of exploitation and rapture on other planets, moons, or asteroids, and in their eyes, the Anthropocene revolution will be permanent. Their vision of the next era in human history will be one where the same murderous emotions that wrecked the Earth will be unleashed on new planets in other parts of galaxies that have Earth-like locations within them.

In subsequent chapters, I will unpack our Earth emotions in more detail. Before I do that, however, I will tell something of my own biography, as it will hopefully help in understanding the source of my Earth emotions and my need to create new ones. I will then introduce my concept of solastalgia for you to consider. I argue that we have now entered the "age of solastalgia," where our emotional compass is pointing in the direction of chronic distress at the loss of loved "homes" and places at all scales. There is already a global pandemic of depression in humans. At its extreme, the lived experience of Earth murder, or "tierracide," is with us. As climate warming and other environmental disasters start to overwhelm residual terranascient places, we will mourn that which is passing.

Only after we resurrect the terranascient emotions will we enter a new era that is compatible with human flourishing, and a whole Earth that is rich, bountiful, and beautiful. We will get to that good place, but first we have to look in the eye the destructive sides of life and human nature. It will not be a comfortable experience. I hope to convince you that Earth emotions are at a point of critical balance right now, a tipping point. The great, ancient drama of creation and destruction is being played out in our minds and before our very eyes. This book is your invitation to participate in the drama and become one of its actors.

Grey Fantail in Duns Creek, New South Wales, Australia.

Photograph by the author.

Chapter 1

A Sumbiography

A Summation of My Green Past

I am at the Donnelly River in South West Western Australia. The place is called "One Tree Bridge." A short walk away, there are the Four Aces, an alignment of four giant Karri trees within a cool, wet forest. As a child I was taken to this place that was so well known to my grandparents. We picnicked there, and the extended family went "marroning" to feast on the monster freshwater crayfish (*Cherax tenuimanus*) that inhabited the crystal and pristine waters of the Donnelly. The birds and wildflowers were all part of the family history of this special place. I remember my grandfather, my Pop Pop, teaching me how to make and light fires for boiling water to cook the marron, and how to cover my eyes and listen for birds before looking for them. He would ask, "What's here?" I recall my grandmother, my Nana, telling my mother about the trees, ferns, and wildflowers, with the child hanging on every word and savoring the thrill of their botanical discoveries. My mother was born into this place of giant trees and abundance, and it was fitting that she be reunited with it after her death. It completed a life cycle that returned her essence to the region of her birth. With my brother, our close cousin, and my wife, I scatter my mother's ashes into the waters.

Before taking you on a psychoterratic (psyche-Earth) journey in search of positive and negative mental and physical landscapes, it is necessary to make the connection between the universal emotions described in the introduction and the context of my own emotional story. The way my emotional states react to the emotional microcosm and macrocosm will be a reflection of my life experiences and how I view biophysical change over time. We all make meaning in our lives, and my psychic alchemy, the mental effort of the creation or transmutation of order from disorder, will be unique.

"Sumbiography" is the term I use to explain the cumulative influences on my life, from childhood to adulthood, that have culminated in my wanting to write this book about the relationship between humans, other forms of life, and nature.[1] These influences include my immediate family and the writers and themes that have come to define my life and my ability to feel, firsthand, the emotional richness of contact with nature. The meaning and importance of the sum total of living together with "nature," people, and other beings is what a sumbiography attempts to describe and acknowledge.

In Australia, such an investigation of the key influences includes acknowledgment of the Aboriginal and colonial histories that predate my entry into this part of the world. These layers of the past stratigraphy of my endemic sense of place are also part of my ecosystem being. They are a given part of my story, and they also set the scene for my "discovery" of solastalgia (chapter 2) as one of many Earth emotions that constitute our lives. The people, places, and relationships that came together to produce me—a bird-loving environmental philosopher trying to understand Earth-human relationships—include my immediate family, naturalists, writers, and friends who together have enabled me to offer a unique perspective on our emotional relationship to the Earth.

Manning and Manjimup

I was born in 1953 into the city of Perth, the most isolated capital city in the world, in the South West of Western Australia (WA). I was also born into Noongar Country, the home of the original people of the Perth area, and the Beeliar or Swan River region. I am the first son of Thelma and Tony Albrecht. Thelma was born in 1928 in Manjimup, "meeting place" or "place of the rushes" in the Indigenous[2] Noongar language, some three hours south of Perth by car. Tony was born in Colombo, Ceylon, in 1926, of Dutch, Portuguese, and Sri Lankan origins. They met at Wooroloo, a sanatorium, both recovering from major operations to collapse a lung as a method of treating tuberculosis. Their lives were ultimately saved by antibiotics, introduced for the first time in WA in the late 1940s, and they were well enough to marry in 1950. Mother was supposed to never have children, but she defied the odds by having me in 1953 and my brother in 1956. The young couple, Thelma and Tony, then set about raising their sons in Manning, a working-class, housing commission suburb south of Perth. Tony was a tram driver and then bus driver for the transport authority. Mother became a housewife and, later in life, a skilled amateur botanist of the native plants of WA.

Manning, in 1953, was a pioneer suburb in many ways. It was being developed into the natural bush on a large piece of land situated near the Canning River, a tributary of the Swan River. Mixed forests of Jarrah (a hardwood tree) and its associated wildflowers were being bulldozed for brick-and-tile Californian bungalows on quarter-acre blocks. Manning was also a "garden suburb" based on the ideas of Ebenezer Howard and the famous Letchworth Garden City model in England. Curved avenues, parks everywhere, halls, playing fields, community health centers, and many more community facilities made Manning unlike the standard Perth suburban development.

As a boy growing up in Manning I knew that the untouched bush was not very far away. Occasionally, we would all be reminded of just how close the bush border lay, as a kangaroo would come bounding down Parsons Avenue. Neighbors would encounter poisonous snakes, and fat, foot-long Bob-tailed skinks (lizards) inhabited everybody's gardens. Poisonous Red-backed Spiders were everywhere, and I do recall that, when I was an older child, a neighbor, Mrs. Hale, was bitten on the bum while on the toilet by one of these poisonous spiders. She survived! Mother and Father, while walking me in a pram in Manning, came across a large Dugite, a local deadly snake, sunning itself on the road in front of them. Manning may have been a garden suburb, but it was also a garden full of deadly spiders and snakes.

I thrived in Manning to become a fast, dark, skinny kid and a nature boy. I was forever in the bush finding lizards and snakes, and in the creeks exploring for jilgies (small marron) and long-necked tortoises. And then there were the birds. I have been in love with birds for as long as I can remember. I was told by my parents that, as a newly graduated toddler, I ran away from home to visit old Mr. and Mrs. Moore three doors down. They owned a venerable Sulphur-crested Cockatoo who lived in a large cage. He was a good talker and could sing "Dance Cocky." Clearly, I was prone to "follow the birds" from a very early age.

Initially, the bird connection to nature was not, however, primarily focused on Manning. It came from my grandparents' house at Dean Mill near Manjimup. It was there, as a little boy, that I came under the influence of my Nana and Pop Pop. Nana was the countrywoman personified with a big voice, ample bosom, and robust body. She could wield an ax as skillfully as giving a grandmother of a hug. Her sponge cakes were to die for. Pop Pop was a "dinky-di" bushman, who worked as a forester among the gigantic Karri, Jarrah, and Tingle trees that cover this part of the lush, wet area of WA.[3] His job was to select the logs that would be cut, dragged, and then trucked to nearby Dean Mill, where they would be assessed, then sawed for timber products. The close contact with the big trees, sawed wood, and smell of sawdust, saws, axes, and hardwood objects gave me a lifelong ardor for the arbor of those southern forests.

My first vivid memory of wildlife at Dean Mill was a lifeless bird, a Silvereye I found in Nana's back garden.[4] It had an undamaged little body and perfect feathers, and I expected it to wake up and fly. I brought it to my Nana, who pronounced it dead, and we buried it. She showed me her caged canaries and budgerigars and promised me I could have a baby budgie for a pet if I was a good boy. Tweetie, my blue budgie of bright thoughts, came back to Perth with me that year. It was the start of a lifelong love affair with birds.

Nana and Pop Pop moved from the village of Dean Mill to a small farm only a few kilometers away. Here they lived in an almost self-sufficient way, with abundant fruit orchards, vegetables, chickens, pigs, and dairy cows. Pop Pop built his very own hardwood sawmill. Nana had taken her caged birds from Dean Mill to the farm, and I recall her pet Pink and Grey Galah (a type of cockatoo).

The creek that ran through the property harbored many poisonous Tiger Snakes. Every so often, a fairly small, thin one had the audacity to come into the farm garden, get though the half-inch-round bird wire into the bird cage, and take one of Nana's precious inmates. However, it was one thing for a snake to get inside the cage and eat a plump budgie or canary; it was quite another to get back out through the bird wire with a bulge in the belly. It would be stuck in wire limbo, unable to get back out. I witnessed my Nana deal out severe punishment when such an infraction was discovered. Out would come the garden shears, and with a few ungrandmotherly words said, the trapped snake would have its head removed. The remains would be solemnly buried . . . the grieving not for the snake, but for the lost canary or budgie. Grief and mourning for the death of the nonhuman was locked into my childhood.

Death or near death for the human was acutely appreciated when I was nearly struck by a dead Tiger Snake that was strung over the hedge at the farm gate. Its fangs missed my arm by millimeters. It must have been a reflex movement, but the headlines very nearly read, "Boy killed by deadly dead snake."

When Pop Pop oversaw the felling of trees in the forest, during the nesting season, baby birds would often spill out of nesting hollows and other nest platforms. It was Nana's task to rehabilitate the baby birds that Pop Pop brought home. Mostly, the orphaned birds were small parrots, Western Rosellas, and they were to be important for me when later I got my first large aviary and Nana gave me a pair of these birds. Her caring for the wild, orphaned baby birds stayed with me as an ethic all my life. It takes hard work, but saving life and allowing it to have a "second chance" was intrinsically valuable. Nana's life motto was "animals and children first." She was way ahead of many who now see nonhuman animals as kin.

The small farm ran as a proto-permaculture paradise, and it flourished through Nana and Pop Pop's diligence and productivity. It seemed an ideal place for a terranascient and good life. It became the blueprint for my lifelong quest to find a home that could replicate such a near sustainable, non-idealistic rural idyll. The condition of self-sufficiency in a beautiful, verdant, and fecund place would later exercise my mind. My idea of a new human era founded on a benign relationship to life and living processes was based on the family farm writ large. Nana and Pop Pop were already living a life I have recently associated with what I call "the Symbiocene" (see chapter 4). It would take the rest of the world some time to catch up to them.

In the Manjimup forests, there was a very special bird that became my bird totem. When I was just a boy this bird would come very close to me, fly around my head, and chatter to me. I thought I must be special for this small grey bird to willingly come so near. It was only in adulthood that I discovered the reason why the Grey Fantail became so intimate. As a member of the flycatcher family, it used me as a mobile human broom to sweep, up into the air, flies and other insects that it could snap up in its acrobatic flight path. It will do this trick for anyone who enters its territory. Despite this hard truth, the Grey Fantail has remained my lifelong totem bird, and I care for it and its habitat. By having close contact with a bird like the Fantail, I developed a sense of kinship to a nonhuman being, something I would carry with me for the rest of my life.

By the time I was ten years old, visits to the farm at Dean Mill would find me heading to the garden, the woodpile, or the bush. But when it rained, or in the evening, Nana had one indoor treat for me that further shaped my whole life. Despite their meager income, Nana received her regular subscription copy of *Animals Magazine*, edited and produced by the Nairobi-based naturalists Armand and Michaela Denis. I devoured every copy she had, many times over. The magazines were in glossy color, and full of stories about the wildlife of Africa and the world. I would disappear in their pages for hours.

I thank the Denises, Gerald Durrell, Jim Corbett, and Jim Kjelgaard for firing up my imagination for animals, birds, and the exhilaration of being in wild places and confronting wild things. Gerald Durrell (*My Family and Other Animals*) also reinforced the value of a sense of humor in leavening the seriousness of the natural world. Jim Corbett (*Man-Eaters of Kumaon*) repeated what my grandparents were telling me: that if you want to know what is going on around you in a forest, the primary sense must be hearing. For someone hunting a man-eating tiger, the warning calls of animals around you could save your life. Jim Kjelgaard (*Lion Hound*) captured the essence of hunting in the wild, and the dangers to dogs and humans when confronted by something as wild and dangerous as a Mountain Lion. His books also gave me a lifelong love of dogs, big dogs. With all three writers, I just loved that they could take the boy reader into the same spaces as bird watching in the depths of Karri forest could take me: the obliteration of the boundary between the knower and the known, of inside and outside, and the disappearance of time. I would later call this state of mind "eutierria," or a good Earth feeling.

I knew, even during that time in my life, that I had to have nature and wildlife close to me. More, I had to be deeply immersed within it. My emotional life was tied to the endemic creatures of Dean Mill, Manjimup, and the Swan River coastal plain. I was driven by what I would later call "endemophilia," the discrete emotion tied to particular special places, and love for that which is unique or endemic to them.

Dark Issues

It was when I was about twelve years old that my father finally discovered that I had over 120 birds in the backyard in various rudimentary cages I had built, plus ponds full of fish and tortoises. He realized too late that

I had become a "bird brain" and that my positive Earth emotions had created a calling for me. My father also played a cataclysmic role in my emotional maturity. I was rocked by his sudden death when I was sixteen. He took his own life in October 1969, and my beloved grandfather died of cancer two weeks later. Together, my mother and younger brother and I had to strike out in a new and independent way to secure our emotional and physical security. My terranascient childhood had been struck an enormous psychic blow.

There was one other "dark" issue that gnawed away at my "ideal" childhood. Being very dark-skinned, I received a lot of racist insults while at school and playing competitive sports. My name at school was "Darky," and my brother, although lighter-skinned than me, was "Little Darky." I was also called "nigger," "abo," and a "boong," a term of abuse with deeply insulting connotations, directed, by some white Australians, toward those considered to be inferior Aboriginal people. My retreat into birds and the wild country around me was, on reflection, a way of avoiding a world where my appearance dictated how I was negatively perceived. I had an early empathy for the Aboriginal people of Western Australia, based on the injustice perpetrated on them by racism. I was a dark kid in white Australia, and that sense of being different has never left me. It explains both my ongoing interest in the Australian Aboriginal worldview and an independence of mind that refuses to comply with white privilege and asserted superiority. I have gentle anarchist tendencies, if anarchism means "order without a (white) ruler."

These emotional upheavals of death and racism, plus my own teenage maturation, saw me switch from reading about birds and other animals to reading and writing poetry. As a way of dealing with the wrenches in life, I also became an avid reader of existentialist philosophy. I had a thirst for knowledge about the human condition, and that saw me collect a library that dealt with the big questions of human existence. In my late teens I combined university studies with laboring and gardening, for both material and psychological sustenance. The gardening and nurturing of plants and animals kept me close to the Earth, and that combination of life and Earth nurturing, and philosophy, was to stay with me for the rest of my life, so much so, that I now call myself a "farmosopher."

Despite the turmoil, my old calling for the study of birds had taken me to the University of Western Australia to become an ornithologist. I failed in meeting this calling, largely because my father's death had irrevocably put me on a new course in life. Moreover, the cultural revolutions of the late

1960s took hold of me and lured me away from endless hours of lectures and practical classes in chemistry, biology, and geology. I became a hippie and was lost to the world of biological science. I had not fallen out of love with birds, but out of favor with the idea that the only career path to a life with birds was to be had via hard science. Moreover, the politics, values, and culture of human social life became overwhelming obsessions. It was out of zoology, and into sociology and philosophy.

I could see how human values, in particular, impacted the natural world in a massive way. The flora and fauna of my home state were disappearing, largely due to active political value decisions. I overtly opposed the wood chip industry in South West WA, as I saw it as an act of despotism against the forest. Although I came from a family of loggers and millers, clear-felling forests and cutting down hardwood trees for tissue paper seemed to me an abomination. My ethics were already tied to my emotional being. I had become an "anarcho-greenie," and I was proud of it.

I actually progressed through university after my first year of absenteeism and failure in science. I completed a degree in social science (sociology), and then commenced a PhD in philosophy, a few years after completing the equivalent of another whole undergraduate degree in philosophy. I had chosen a topic that examined the history of ideas related to the concept of organicism. The idea that society is, or is like, a natural organism was the key theme that I pursued, through the writings of the ancient Greeks, the German Idealists, Karl Marx, and the British Idealists. Hegel's organic and dialectical philosophy figured prominently in my studies, and I had a lovely ethicist and Hegelian scholar, Julius Kovesi, supervising me at the University of Western Australia. I could see connections between the older tradition of organicism and its reemergence in the new environmental philosophies of what was then the late twentieth century. Hegel meets Gaia! I was making slow but steady progress when another death cut short my intellectual studies. Julius had a heart attack and could not continue his work. He died not long after.

I was philosophically completely lost, as Hegelian supervisors in a British Empiricist philosophy department were very rare birds. If I was to continue my PhD, I would have to find a new supervisor, so I investigated the other universities in Perth. No luck. After an exhaustive search, I discovered that the only living Hegelian philosopher in the southern hemisphere, Dr. Bill Doniela, worked at the University of Newcastle on the east coast of Australia. With a spirit of adventure and a new start, we decided to pack up and go east.

A New Sense of Place

After two decades of being Western Australians, with my wife, Jillian, and stepdaughter, Sarah, we put our possessions and two large dogs into an old Ford station wagon and headed off to Newcastle four thousand kilometers away. Even as we drove east, a sense of nostalgia for the West permeated our emotions. In Newcastle we found a new university, home, employment, and ecosystem to immerse ourselves in and understand. Yes, there was nostalgic melancholia for a while, as we missed both the landscape and people of the West, but soon the delights of the Hunter Region began to open up for us.

The native birds of the area were more numerous and, in many respects, more brilliant than those of the West. Newcastle was also at the confluence of cool temperate and tropical climates, and in the summer, this borderline was exemplified by butterflies of all types and colors. Iridescent Blue Triangle beauties would glint in our eyes while big Orchard Swallowtails would sail around the garden searching for succulent citrus buds for their eggs, then caterpillars. A new sense of place was developing, one that had echoes of the previous place in Perth, but with new layers of richness that New South Wales had to offer. My bird totem, the Grey Fantail, had become rare in suburban Perth, yet the parks and gardens of Newcastle and its environs seemed to be its stronghold. I soon felt at one with my new home and sense of place, as my new love of place was being built by my biophilia, or love of life, and an emergent endemophilia.

I became involved in the life of the university, and I gradually completed my PhD. My relationship to birds was once again ignited as I made new ornithological discoveries and met with others who shared my orniphilia. From the love of birds, it was an easy step to work with like-minded people, such as the late professor Max Maddock, to protect remnant wild places around Newcastle where birds were to be found. For ten years of my life I took on volunteer roles I had never before contemplated, such as becoming the foundation secretary of the Hunter Wetlands Trust, and a member of the board of directors of its associated Wetlands Centre at Shortland in Newcastle. Not only was I researching and watching birds, I was creating and protecting habitat for them by collaborating with other citizens to achieve a common valued outcome. About thirty years later I would call this working together with others to protect loved places "soliphilia." It had its origins in my early experiences of the necessity for cross-factional political environmentalism, for the protection and creation of wetland ecosystems at Newcastle. You have to love something before the motivation to protect it becomes paramount, and that love overrides political affiliations.

I also became interested in the history of Australian ornithology through the work of John and Elizabeth Gould, the British pioneers of high-quality folio books of bird illustrations. It was a surprise discovery to find out that they had traveled to Australia in 1839–40 and spent time in the Hunter Valley. We had purchased an original lithograph by John Gould and, by chance, on the folio page of notes that came with the image of the beautiful black and gold Regent Bowerbird was an account of John Gould collecting and observing this bird at the mouth of the Hunter River at Newcastle.

I immersed myself in their published works and built a picture of what they had witnessed while in the Hunter Valley. Jill and I gave a presentation to the Royal Australasian Ornithologists Union Eighty-Sixth Congress in 1988, where I read the words of John Gould, and Jill those of Elizabeth Gould, in a talk and slide presentation. Building on that material, a slender monograph, *The Goulds in the Hunter Region of N.S.W. 1839–1840*, was the outcome of the research work we completed.[5] In the book, we not only cataloged the birds seen, collected, and drawn by the Goulds, but also noted the vivid description of the landscapes at that time. The Goulds were more than generous in their descriptions and appreciation of the "beautiful parklike scenery" of the Hunter Valley.[6]

In addition to gaining a historic vision of the Hunter Valley through the eyes of John and Elizabeth Gould, we concentrated our travels on remote and beautiful places. The Hunter Valley has wetlands and heaths along the coast, but farther inland it has remnant rainforest that once was the home of the red cedar, so highly prized for its lustrous red timber. There is also alpine country, with the last remaining stands of Antarctic Beech trees (*Nothofagus moorei*) from ancient Gondwana connections to South America. The whole region filled me with what the geographer Yi-Fu Tuan called "topophilia," or love of place.[7] Given my knowledge of the history of the region and direct experience of its beauty, I was able to seamlessly transfer my positive Earth emotions nurtured in Western Australia right into the Hunter Valley. I had acquired a new and very positive version of what George Seddon called "sense of place."[8]

At the same time as my head was full of positive historical images of the Upper Hunter—described by some as "the Tuscany of the South," called "beautiful" and "parklike" by Elizabeth Gould, and "a pretty place far beyond what I had anticipated" by John Gould—I had my own confrontation with the negation of all of my preconceived images of the Hunter Valley. On one journey into what I hoped would be "Gould country," I was confronted by a scene that captured the full impact of the desolation, by massive terraphthoric forces, of the Upper Hunter Valley.

The New England Highway runs through the center of the Hunter River valley, and the coal-mining region starts at the town of Singleton. As you leave Singleton, there is a steep hill that takes you to a point where, at the rise, a vista of the whole valley opens up. I stopped and got out of my car and looked west up the valley at the desolation of this once beautiful place. In front of me lay hundreds of square kilometers of open-cast black coal mines feeding two large power stations and a rail system to take coal to the port of Newcastle for export by ship. The air was thick with dust and there was the acrid smell of burning coal from the spontaneous combustion that was erupting in various actively mined locations.

The massive draglines removed bucketloads of earth and filled equally massive dump trucks with the "spoil." There were flat areas, pockmarked with the sites where explosives had been placed for the next blast to remove the "overburden." A dull roar went up in the far distance as a mine detonated a panel of overburden, and a cloud of orange smoke drifted over that end of the valley. Blast dust pollution mixed with that of the two huge power stations, Bayswater and Liddell, which sit on the central valley floor. The power stations were ejecting huge amounts of steam from their cooling towers, creating their own clouds on an otherwise cloudless day over the valley. In addition to the steam, carbon dioxide and toxic but invisible fumes left the giant chimney stacks in amounts measurable in thousands of tons per annum. Running parallel to the road was the railway, taking the washed coal to the coal loaders at the Port of Newcastle, and then by ship to the black coal coke ovens and power stations of the world. The conjoined diesel engine trains spewed a plume of black particulates into the air, as they pushed their heavy load out of the valley and toward the coast.

I could see that the creeks and the Hunter River itself were being seriously degraded by coal dust and the products of disturbed land and erosive forces. Not only had agriculture cleared the alluvial land close to the river, now the coal mines and their infrastructure were polluting the streams and creeks. I could see riparian vegetation dying or in ill health everywhere. The industrial landscape below me was the very opposite of the memories of the Goulds and their description of the rainforests and the undulating hills of Tuscany. It had become the workplace of terraphthorans, terraforming the valley into a series of flat-topped mesas and voids full of toxic mining waste. I stood there, taking in the full impact of what I would later call "tierratrauma," or the direct experience of an acute traumatic event of environmental destruction. In the Hunter Valley of New South Wales (NSW), there is an excellent illustration of human dominance over the biophysical environment. If you wish to understand what the Anthropocene, or period of human dominance over all nature, means, then this sacrifice zone is where you must come.

Later, back at my university, I began to think deeply about how to respond to the shock to my naive sense of place. I had been living a fantasy, through the words of pre-desolation pioneers. But I had just witnessed landscape destruction that must rate as one of the largest expanses of open-cast mining on the planet, and I needed to respond. At the same time, citizens from the Upper Hunter began to communicate to me their concerns about the sheer scale of desolation taking place in their beloved Hunter Valley. They rang me at work, clearly upset, as they told their stories of the chronic assault on their sense of place. Mine by mine, dragline by dragline, these people were being driven insane by forces beyond their control. My emotional reaction resonated with theirs. We were not experiencing different responses; our emotions were telling us that both the Earth and the human psyche were in trouble in this place. I knew then that there must be a concept, a word, for that negative, distressing emotion, and I set about to understand what it might be. "Solastalgia" was about to enter the world from the Upper Hunter Valley of NSW.

My life history, and the themes that run through it, explain to some extent my ability to feel, first-hand, the emotional richness of contact with nature. It also provides the opposite possibility, in the form of severe distress at sickness and death in nature. From solastalgia onward, I have been working away at describing and defining what I call the "psychoterratic typology," or all the positive and negative Earth emotions that I have found in the literature, or that I have created. It was not enough to define solastalgia: it must also have many closely related positive and negative concepts in the psychoterratic "family."

I realized, too, that our Earth emotions sit within a much larger narrative about nature and life. This book is the first attempt at creating the full context for these emotions and the feelings that we all must have experienced but have no proper home for their expression. My home, as I write this book, is Wallaby Farm in rural New South Wales. I am surrounded by birds, and there are many self-sufficient echoes of Manjimup and Dean Mill at this place. Although Wallaby Farm is already being hammered by climate change, it is my hope that my sumbiography will end at the point when humans have halted the ascendancy of terraphthoran forces and have allowed the terranascient within us all to give birth to the Symbiocene. I trust that by the time my life has finished, children will be able, once again, to grow up in "my" world, and it will be an era in Earth history that gives generously of positive Earth emotions. My life cycle, like my mother's, would then be complete.

Finally, it is also my hope that the idea of a sumbiography will be useful to others who wish to understand their own response to the challenges faced in this century. Environmental educators, for example, could use this way of revealing values and emotions as a teaching aid for student evaluation of the past and their construction of visions of the future.

Coal trains in the Hunter Valley, New South Wales, Australia.

Photograph by the author.

Chapter 2

Solastalgia

The Homesickness You Have at Home

I hold within me a strong sense of place. My emotional compass, my inherited biophilia, instinctively directs me to beautiful and biologically rich environments where I want to be "in place" with as much place attachment as possible. Hence, to directly experience the full destructive force of coal mines and power industries on the people and place identity of the Upper Hunter region was a personal psychic upheaval for me. I also knew that place identity was important for rural people, and that rural landscapes were traditionally seen as places of great beauty, with strong emotional attachment running deep in family history. The word that came closest to describing that positive feeling is Yi-Fu Tuan's idea of "topophilia," or love of place.[1] The *topos* (place or region) was an object that humans could love. However, it becomes hard for that love to be tightly held when the land is being blown up, scoured by a giant excavator, then dumped in a truck, train, or ship. It did not take too much experience of that desolation to generate within me strong, empathetic ties to these rural people under assault.

For the Indigenous people of the region, the new land incursion by coal mining must have seemed like a repeat of the first wave of invasive colonization in the early nineteenth century. In a sense, they were already prepared for the insults to the Earth and people associated with the burning of coal.

Close neighbors of the Wonnarua people of the Upper Hunter, the Awabakal people of Newcastle, at the mouth of the Hunter River, have a Dreaming story that warns of the dangers of burning coal.[2] Their story says that coal was originally on the surface of the Earth, but that the ancestors covered and buried it. They say that there was a time when a great darkness came over the land and blotted out the sun. The darkness originated from a fire, burning from coal, deep within the Earth, and issuing acrid, black smoke from a hole at the surface. The elders decided that, in order to bring back the light, the darkness had to be stopped from escaping the hole. All the people gathered rocks, trees, and plants, covered the hole, and put out the fire. They also covered coal wherever they found it on the surface of the Earth. Then as thousands of generations of people walked over their country, the coal was rendered safe, as it was compressed and remained deep underground.

In many places in the Hunter Valley, the Permian rocks are compressed shale with countless fossil impressions of an ancient (extinct) plant genus called Glossopteris visible on exposed surfaces. Western scientific geology, evidence of fossil plants, and the Dreaming seem not that far apart. Perhaps the ancestors also had a glimpse of the link between burning coal, loss of health, species extinctions, and climate change. We need to leave the world's remaining coal reserves in the ground in order to avoid gross air pollution, such as that experienced in the industrial cities of the world, and to limit climate warming to less than two degrees Celsius. The Awabakal made such a connection thousands of years ago.

While the Wonnarua were dispossessed from their own lands and witnessed wholesale land clearing that included the ringbarking and burning of riparian rainforests, the land and soil were still in situ, now growing maize and cows rather than red cedar, kangaroos, and emus. For the descendants of the Indigenous people, colonization never ended; in fact it had reached new heights of land and psychic desolation under the brutal impact of open-cast coal mining.

For the dairy farmers, viticulturists, croppers, and horse breeders, the new coal industry had emerged from the older traditions of underground mining, with a pithead and a surrounding village that housed the miners and their families. That kind of relatively low-impact underground mining was typical of the Lower Hunter around Newcastle. However, as the easily accessible coal seams ran out, and underground mining became uneconomic in that part of the Hunter, the next big resource for bulk coal would be the deep strata of black coal in the 240-million-year-old Permian geology of the Upper Hunter Valley. The supply chain needed for coal power generation in New South Wales and for the global export market was heading inland from the city of Newcastle and its mining suburbs such as New Lambton and Wallsend. There, in the Upper Hunter Valley, a good hour or more inland from the coast, and centered on the towns of Singleton and Muswellbrook, grew new megasize open-cast mines, large coal-fired power stations, and the infrastructure, such as railways, to service the flow of coal.

The farming-orientated economy began to turn toward a new order, based on the wealth and employment delivered by the black gold. Some farmers wanted to stay in farming and continue to work the land, while others, particularly those in the failing dairy industry, could see that jobs in coal mining were being offered, just as neoliberalism was ensuring that the dairy industry was about to be deregulated out of existence. Clearly, farmers who became economically dependent on the mines for their income had mixed feelings about this new land use, but they had become hooked by the money and job security. There was even a name, the "golden handcuffs," given to those who were heavily dependent on employment in the mines, because once committed to the big salaries on offer, people would become dependent on them to maintain high living standards and pay off the mortgage, car, and boat. Even though some might have had misgivings about the assault on their valley, by the mining and air pollution from the power stations, they could not publicly criticize their own employers, nor could they support neighbors who wanted to continue farming. The economy of the region had become inextricably chained to coal.

The farmers who remained on their properties had no such divided allegiances. They lived close to the land and the weather. They were acutely receptive to big vistas, the brilliance of the Milky Way above their properties at night, abundant wildlife—from micro bats to kangaroos—prolific birdlife, and the pure, sweet scent of petrichor after much-needed rain. As every element of what they valued in rural life came under assault, the changes to their place negatively affected their sense of place. Although affected differently, both the Indigenous people and the colonists suffered from a condition that was close to the traditionally defined concept of nostalgia.

Nostalgia

It was a Swiss trainee medical doctor, Johannes Hofer, who created the concept of nostalgia, in 1688, and published its detailed diagnosis in a dissertation written in Latin in Basel.[3] Nostalgia was a translation into Greek and New Latin of the German word *heimweh,* or the pain for home, most comfortably translated into English as "homesickness." The term "nostalgia" (from the Greek *nostos*—return to home or native land—and the New Latin suffix *algia*—suffering, pain, or sickness from the Greek root *algos*) was the sickness caused by the intense desire to return home when away from it. It could be a profound cause of interrelated mental and physical distress.

Nostalgia, defined as an intense desire to return home, was considered to be a medically diagnosable psycho-physiological disease right up to the middle of the twentieth century. According to Hofer, the symptoms of nostalgia included a whole range of psychological and bodily afflictions, ranging from intense sadness to palpitations of the heart. He suggested that nostalgia was most likely to be experienced by people who were forcedly or deliberately removed from their home environment, such as soldiers transported to fight on foreign soil. In addition to war, other forms of prolonged absence from loved home environments were a likely cause of nostalgia.

Found in the English language from the mid-eighteenth century, "nostalgia" gradually became associated with a sense of regret and longing for places or periods in the past where a person felt more at ease. The more frequent modern use of the term loses its connection to the geographical or spatial "home" and suggests a temporal dimension, or a "looking back," a sentimental desire to be connected with a positively perceived period in the past.[4] In most academic instances, nostalgia is no longer discussed in psychological or medical literature as a diagnosable illness signifying deep place-based distress.

However, as I have argued elsewhere, there is still a case for the relevance of Hoferian nostalgia in contexts where people closely tied to traditionally occupied homelands are forcibly removed, or have no say in the matter of being displaced.[5] The impact of forced relocation of indigenous people the world over reverberates as a wave of "sickness" through their cultures to the present day.

In parts of sub-Saharan Africa and the Middle East, and more recently Myanmar, both climate change and religious, political, and ethnic persecution are likely to cause traditionally defined homesickness in the refugees who flee or are pushed into temporary accommodation such as refugee camps. Populations who are removed from areas designated to become large industrial developments or dams are also subject to forced removal by government fiat. Nostalgia, development, and environmental desolation often go hand in hand.

Other researchers of place identified the problem of once-loved places that were turning toxic for those who continue to live within them. Edward

Casey, in *Getting Back into Place*, forwards the idea of "place pathology" to explain what happens when a place becomes unhealthy.[6] In the context of North America, he has said:

> It is a disconcerting fact that, besides nostalgia, still other symptoms of place pathology in present Western culture are strikingly similar to those of the Navajo: disorientation, memory loss, homelessness, depression, and various modes of estrangement from self and others. In particular, the sufferings of many contemporary Americans that follow from the lack of satisfactory emplacement uncannily resemble (albeit in lesser degree) those of displaced native Americans, whom European Americans displaced in the first place. These natives have lost their land; those of us who are non-natives have lost our place.[7]

I was struck by this analysis of place, because it applied equally well to the experience of Indigenous people in Australia and Papua New Guinea. The anthropologist W. E. H. Stanner, writing about the plight of Indigenous people in Australia and New Guinea during the colonization process of the nineteenth and twentieth centuries, described a similar kind of syndrome based on the cumulative distress linked to homelessness, powerlessness, poverty, and confusion. He wrote:

> What I describe as "homelessness," then, means that Aborigines faced a kind of vertigo in living. They had no stable base of life; every personal affiliation was lamed; every group structure was put out of kilter; no social network had a point of fixture left. . . . In New Guinea, some of the cargo-cultists used to speak of "head-he-go-round-men" and "belly-don't-know-men." They were referring to a kind of spinning nausea into which they were flung by a world which seemed to have gone off its bearings. I think that something like that may well have affected many of the homeless Aborigines.[8]

While I could see that "place pathology" was a useful term to explain the plight of people in mining-affected communities, Casey was mainly focused on people who were displaced and could not "re-enter" lost places. In the Hunter Valley, there were already many farms and villages that had been lost in this way to mining. You simply cannot live next door to an active coal mine, and many were forced to sell to the company and leave their farms. These people became displaced and were forced to go elsewhere to become re-emplaced. Many writers and researchers examined this process of dis-emplacement and its impact on the emotional life of people, but very few focused on places that were not completely lost, nor strictly pathological in the sense of causing disease, but were impacted by forces that made the people living within them feel distressed, dis-eased, or ill at ease. I was looking at a place-based, mainly existential problem, one not adequately covered by my reading of the literature prior to 2003. Also, my focus was not solely on the nuances of place, but on the active relationship between human emotion and the biophysical state of a given place. The negative transformation of a loved place triggers a negative emotion in the whole person who is still emplaced. Their love of their place remains, but they want back those positive elements of place that gave them such a positive sense of place prior to the "invasion."

The people in the Hunter Valley I was focused on were still "at home," but were feeling a melancholia similar to that caused by traditional nostalgia related to the breakdown of the normal relationship between their psychic and emotional identity and their home. These people were losing the solace or comfort once derived from their relationship to a home that was now being desolated by forces beyond their control. In the Upper Hunter, people were suffering from both imposed place transformation by the mining industry, and powerlessness in the face of environmental injustice, perpetrated on them by transnational mining companies and the state government of NSW. To oppose the negative changes to their home environments, they needed to challenge some of the most powerful institutions on the planet: mining multinationals and the full resources of state and federal governments.

Other Influences

Having taught the sociology of health and illness during my early university teaching career, I had been thinking about the relationship between ecosystem distress and human distress for some time, well before my own direct experience of the distress in people affected by mining in the Hunter Region of NSW. The breakdown of a healthy connection between general human health and the health of the biophysical support environment, or home at various scales, had been considered by many thinkers from the Greek physician Hippocrates (ca. 400 BC) onward.[9] Hippocrates, for example, saw a close connection between a healthy or unhealthy environment and a healthy or unhealthy human being. Before any close diagnosis of a particular person's state of health, he first examined the general relationship between the climate and biophysical features of a region, the cultural values and occupations, and the distinctive health profile of its human inhabitants. He observed:

> Whoever wishes to investigate medicine properly, should proceed thus: in the first place to consider the seasons of the year . . . then the winds . . . the quality of the waters. In the same manner, when one comes into a city to which he is a stranger, he ought to consider its situation, how it lies to the winds and the rising of the sun. . . .
>
> From these things he must proceed to investigate everything else. For if one knows all these things well . . . he cannot miss knowing . . . either the diseases peculiar to the place, or the particular nature of common diseases, so that he will be in no doubt as to the treatment of the diseases.[10]

Hippocrates would have found the Hunter Valley to be a good case study in how change to a biophysical region had impinged on the physical and mental health of its people. He also had pristine environments and villages close by that would have served as comparisons to the unhealthy transformations of the Upper Hunter. Hippocrates provided a sound foundation for ecologically informed public health.

The pioneering sociologist Émile Durkheim also would have found the Upper Hunter region to be a rich case study in his concept of "anomie," or social normlessness, and its impact on the mental health of people.[11] As social order is demolished, in the example of farming communities tied to a particular type of biophysical order in the farming landscape, people lack a sense of direction and purpose that make life well worth living. The path from anomie to anomic suicide is often surprisingly short.

In a similar vein, the existentialist philosopher Albert Camus also made the connection between suicide and issues such as the desire to end meaninglessness and purposelessness in life. Camus saw the undermining of the goals and purpose of life as being at the core of recognizing its absurdity and the anguish that can follow such an existential state. In this, Camus was following Nietzsche, who was one of the first to articulate the hugely emotional connection between home and the heart. Camus, through the Nietzschean rebel, lamented:

> From the moment that man believes neither in God nor immortal life, he becomes "responsible for everything alive, for everything that, born of suffering, is condemned to suffer from life." It is to himself, and to himself

alone, that he returns in order to find law and order. Then the time of exile begins, the endless search for justification, the nostalgia without aim, "the most painful, the most heart-breaking question, that of the heart which asked itself: where can I feel at home?"[12]

When the limits of one's world, its laws and order, are destroyed by forces beyond one's control, then home becomes not only toxic, it becomes "nostalgia without aim." Although perhaps not in the same grand category as God or immortal life, failure to be able to believe in the certainty that a rural "home" will even exist in the near future is, in my view, sufficient cause for existential distress. As was indicated above, with the Navaho and Australian Aborigines, once their world was collapsed in rounds of warfare, then dispossession and relocation for the survivors, existential crisis followed.[13] In all cases, as the home territory is either dramatically changed or people are forcedly relocated to another "alien" land, they will suffer place-based existential distress. In the Australian context, Deborah Bird Rose captures the essence of what being in place and displacement mean, in an account of what the term "country" means to Indigenous people:

> Country is not a generalised or undifferentiated type of place, such as one might indicate with terms like "spending a day in the country" or "going up the country." Rather, country is a living entity with a yesterday, today and tomorrow, with a consciousness, and a will toward life. Because of this richness, country is home, and peace; nourishment for body, mind and spirit; heart's ease.[14]

The Aboriginal people of Australia had found a solution to what great thinkers like Nietzsche and Camus saw as one of the biggest questions on Earth. Coming to intimately know a place as home is at the same time a way of achieving heart's ease.

Ecosystem Health

As I moved from social considerations of public mental health in general to teaching within the context of environmental studies, I came under the influence of two twentieth-century pioneers and champions of relationships between land health and human health, the US environmental thinker Aldo Leopold, with his concept of "land health," and Australia's own early environmental thinker Elyne Mitchell.

Aldo Leopold wrote about a land ethic in *A Sand County Almanac* (1949), which not only broke new ground in the emergent domain of environmental ethics, but also created the concept of "land health."[15] Leopold defined this new idea as "the capacity of the land for self-renewal." He even wrote about "sick landscapes," and the failure of landowners to see their culpability for the land morbidity in their possession.

However, before Leopold's ideas on land health were posthumously published, a little-known Australian farmer named Elyne Mitchell, in her book *Soil and Civilization* (1946), was attempting to explain to her fellow Australians the importance of the connection between human mental health and ecosystem health, as exemplified by the state of the soil. In the context of the impoverishment of the Australian environment by agricultural activity she wrote:

> Often it seems that genius does not occur where there is perfect physical health. But no time or nation will produce genius if there is a steady decline away from the integral unity of man and the earth. The break in this unity is swiftly apparent in the lack of "wholeness" in the individual person. Divorced from his roots, man loses his psychic stability.[16]

Mitchell had in mind the chronic loss of topsoil in Australian agriculture and forms of land clearing and overstocking that were desolating what she described as the "essential Australia." However, her identification of a loss of psychic stability with the degraded state of the environment was, for me, a very important discovery. Replace "chronic lack of land care" in farming and forestry with open-cast coal mining, and we are in much the same situation as with the people of the Upper Hunter and their Earth-based distress.

A similar relationship, between a state of the environment and states of human mental health, was expressed in the work of the transdisciplinary thinker Gregory Bateson in the 1970s. He argued, in the context of the Great Lakes of North America, that

> you decide that you want to get rid of the by-products of human life and that Lake Erie will be a good place to put them. You forget that the eco-mental system called Lake Erie is a part of *your* wider eco-mental system—and that if Lake Erie is driven insane, its insanity is incorporated in the larger system of *your* thought and experience.[17]

Despite Bateson's prescient insight into the centrality of eco-mental relationships for human mental health, they remained poorly articulated in the general literature. Finally, I was influenced by David Rapport's concept of "ecosystem distress syndrome" and realized that there must be a human eco-mental distress syndrome that corresponded with biophysical ecosystem distress.[18] I set out to find out what it was.

I was also aware that other cultures and languages had concepts similar to the one I was searching for in English. For example, the Inuit of the Arctic have the word "uggianaqtuq" (pronounced OOG-gi-a-nak-took), which has connotations of a "friend acting strangely" or unpredictable behavior. They have applied this word to the way climate change is impacting their culture. They were forced to take a concept from their social context and put it into an environmental one, in order to make sense of the novel changes to their "home." Hopi Indians used the word "koyaanisqqatsi" to describe a state of life that is out of balance and disintegrating.[19] I felt these important concepts needed an equivalent word in the English language.

As a philosopher, I was aware that great psychological and philosophical thinkers, such as Freud and Heidegger, had already identified a type of anxiety or dread associated with the opposite of being at home. Freud's notion of *unheimlich*, translated in English as "uncanny," refers to something sinister or threatening within "the home," understood both as the seat of consciousness and as the place where one lives. That which should be a source of comfort flips into something threatening, making the occupant feel uneasy and anxious. Something familiar and well established in the mind becomes "uncanny" after a process of repression turns it into an alienated mental construct. The familiar comes under threat from elements within the unconscious, undermining its previous safe conscious state.[20]

Heidegger's concept of *Unheimlichkeit* is also translated "unhomeliness" or "uncanniness," and it, like Freud's *unheimlich*, refers mainly to an uncertain state of being rather than a concrete place-based emotion.[21] According to Heidegger, the topos is, at one level of reading, not the external landscape: it is the "shape" of the contents of consciousness. However, it is clear that extensive mining in a once rural landscape renders the total landscape as unfamiliar and uncanny. As objects and beings within the landscape are removed by mining, that which once delivered the comforts of home becomes desolated. Open-cast coal mining generates "unhomeliness" by the bucketful.

Despite having the concepts of place pathology, psychic instability, the uncanny, nostalgia, and eco-mental distress syndrome, I was not happy with past attempts to understand and explain the link between failing ecosystem health and human mental health. My task seemed more psychological than philosophical or medical. It seemed a very simple situation to analyze. A huge and powerful transformational force (mining) had entered the place and lives of rural people as a kind of invasion, and it was distressing for them.

It was clear to me that the English language had no term within it to describe what I had felt while overlooking the Hunter Valley, and what I sensed was being conveyed to me by the citizens who rang me at work to express their emotional turmoil about what was happening to them. I needed a new concept, a "canny" one, which could be readily understood by ordinary people out in the world of land degradation, coal mining, and power stations. Such a word might even be useful in academia.

I thought we needed, in English, the idea of a place-based emotion that captures the feeling of distress when an external force, one that we are powerless to prevent, enters the biophysical location or "life-space" within which one lives out a life (the private home, the property, and the region) and chronically desolates it. The place becomes literally toxic, and at the same time one's sense of place becomes negative.

It is not a case of "placelessness," as people are still firmly emplaced within their "home."[22] The feeling has nothing to do with the uncanny as a repression anxiety, or a subtle point about the topology of the mind. It is in your face, existential, raw, and Earthly. And, as this experience is not typically acute or sudden, the existing psychological and psychiatric syndrome called post-traumatic stress disorder (PTSD) does not apply, as the immediate trauma is normally associated with the sudden onset of the causal agent. Finally, traditionally defined nostalgia could not apply, as the people I was focused on in the Upper Hunter had not left their home environment. I needed a single word to capture this distinctive form of human distress caused by place distress.[23]

The Creation of Solastalgia

After the investigation of nostalgia, I felt that I was close to identifying the concept that I was looking for. It had to be a concept that embraced an emplaced, chronic, and painful emotion in the face of negatively experienced environmental change, and I did not want to simply add "eco" in front of some existing form of distress. With the help of Jill Albrecht, I explored the pairings of words with the *algia*, pain or distress, that was at the heart of this issue. Two concepts seemed to me to be very close to what I was searching for: solace and desolation.

The word "solace" is derived from the Latin verb *solari* (noun *solacium* or *solatium*), with meanings connected to the alleviation or relief of distress, or to the provision of comfort or consolation in the face of distressing events. In this dual connection, solace has meanings for both psychological and physical contexts. One emphasis refers to the comfort one is given in difficult times (consolation), while another refers to that which gives comfort or strength. A person, or a landscape, might give solace, strength, or support to other people. Special environments might provide solace for those looking for consolation. If a person lacks solace, then they are distressed and in need of consolation. If a person seeks solace or solitude in a much-loved place that is being desolated, then they will suffer distress. Solace is what provides "heart's ease"; it soothes the disturbed mind and brings that which was discordant back to harmony.

Similarly, the word "desolation" has its origins in the Latin *desolare*, with meanings connected to devastation, deprivation of comfort, abandonment, and loneliness (to be solitary or left alone). It also has meanings that relate to both psychological and physical contexts—a personal feeling of abandonment (isolation), and to a landscape or structure that has been devastated.

In addition, I wanted a concept that had a ghost reference or structural similarity to the concept of nostalgia or homesickness, thereby ensuring historical continuity and affinity with that concept, and one in which a place reference remained imbedded. All the elements were present with the "sol" of solace and the "sol" of desolation, plus the reference to "home" that "nostos" in "nostalgia" provided. That this new concept was an "algia" was abundantly clear to me right from the start.

"Solastalgia" rolled out of my mind and off my tongue as if it had always been there. In the instant I made "sol" connect with "algia," I knew that this new concept would work for me in the way I wanted and intended. It looked right and it sounded right. I had a gut feeling that it would be understood and appreciated by those affected by coal mining in the Hunter Region. Hence the word has its origins in the New Latin word "nostalgia" (and its Greek roots *nostos* and *algos*), yet it is based on two Latin roots, "solace" and "desolation," with a New Latin suffix, *algia* or pain, to complete its meaning. In essence, although a combination of words from Greek and Latin, "solastalgia" is primarily a New Latin neologism.

I define "solastalgia" as the pain or distress caused by the ongoing loss of solace and the sense of desolation connected to the present state of one's home and territory. It is the existential and lived experience of negative environmental change, manifest as an attack on one's sense of place. It is characteristically a chronic condition, tied to the gradual erosion of identity created by the sense of belonging to a particular loved place and a feeling of distress, or psychological desolation, about its unwanted transformation. In direct contrast to the dislocated spatial dimensions of traditionally defined nostalgia, solastalgia is the homesickness you have when you are still located within your home environment.

I have always tied the definition of the emplaced, existential feeling of solastalgia to the state of the biophysical environment. In the case of the Upper Hunter, the clearly defined biophysical entity was the valley, its river, and the landscape contained within it. However, even in the first published journal essay on this new term, I was aware that the physical environment contained many elements:

> The factors that cause solastalgia can be both natural and artificial. Drought, fire and flood can cause solastalgia, as can war, terrorism, land clearing, mining, rapid institutional change and the gentrification of older parts of cities. I claim that the concept has universal relevance in any context where there is the direct experience of transformation or destruction of the physical environment (home) by forces that undermine a personal and community sense of identity and control. Loss of place leads to loss of sense of place experienced as the condition of solastalgia.[24]

After writing these words, I have come to appreciate that I needed to stress the chronic nature of biophysical change, even in situations like fire and flood. My emphasis was not, for example, on the actual fire or flood event and its immediate psychological effects. My focus was the ongoing impact of the changed environment on those who remained in the area affected by the disaster. In addition, as some wanted to include humans as part of their physical environment, I have been careful to define solastalgia as connected to the biophysical components of either natural environments or "built" environments, those constructed by humans.[25]

Fortunately, these willful attempts to redefine solastalgia have been few and far between. For the most part, there has been widespread acceptance of the concept and applications entirely appropriate to the spirit in which I created the term. However, there has been one ongoing source of discontent. My first publication on solastalgia had in its subtitle that it was "a new concept in human health and identity." By this I meant to convey the notion that solastalgia was broadly related to mental health but was not a biomedically defined mental illness. That others have criticized the concept for being potentially biomedically applied is owing in some part to my naivete as a transdisciplinary philosopher in thinking that the use of terms like "dis-ease" (lack of ease) and "illness" (not normal) would not be confusing or misleading to a general readership.[26] Plus I first published in a journal called *Philosophy Activism Nature*, not a journal called "New Discoveries in Biomedical Psychiatric Science." I was wrong. I have always thought that solastalgia can be a precursor to serious and diagnosable forms of mental conditions, such as depression, yet I have never thought that it is a condition that can be medically and psychiatrically diagnosed. It is a condition of existence, an emotion, not a lesion in the brain.

In addition, in an article on solastalgia and climate change by the journalist Clive Thompson that appeared in *Wired* in 2007, there was an element of confusion about the coauthored work done by collaborating psychiatric and medical researchers on drought and elevated suicide rates among males in rural Australia and the work done by me and nonbiomedical colleagues on existential solastalgia in mining-affected communities.[27] Unfortunately, I was given no opportunity to correct that error before publication, so I have had to live with the possible implication that I endorse the idea of solastalgia as a biomental illness that can lead to outcomes such as suicide.[28]

In an interview conducted by the journalist Sanjay Khanna in 2009, I argued, in response to the query whether solastalgia could be added to the *Diagnostic and Statistical Manual of Mental Disorders* (DSM), that "given that key aspects of solastalgia are existential, the traditions of environmental philosophy and medical psychiatry may not come together so harmoniously. The melancholia of solastalgia is not the same as clinical depression, but it may well be a precursor to serious psychic disturbance."[29] The use of the term "mental illness," with biomedical implications, in connection to solastalgia is not something I can control, but in all my publications I try to indicate the existential, emotional, and human-psychological nature of the concept.

I am also aware that the existential experience underlying solastalgia is not entirely new, only that it is newly defined in English (but possibly represented in many other languages). The experience of solastalgia is likely to be ancient and ubiquitous. What is new, though, is the fact that, in the last half century, solastalgia has become a much more widely felt emotion, under the impacts of ecosystem distress and climate chaos, during the period of rapid growth in what is now being called the Anthropocene. I have suggested that "the age of solastalgia" is likely to emerge during this period of massive change.[30] The global nature of the chronic stressor of detrimental climate change is a novel solastalgic event for all cultures. I hope I am wrong, but solastalgia looks like it is here to stay for a while.

The Global Context of Solastalgia

As further evidence that solastalgia captured a unique aspect of human emotional experience, I have been heartened by the uptake of the concept since its creation in 2003. In academic circles, art, and the public domain, the concept of solastalgia has had considerable international impact and has helped focus transdisciplinary interest in the relationships between humans and place at all scales.

In response to the impact of changing home environments, and to my publications, as well as those of my colleagues, other academics in disciplines as diverse as psychology, psychiatry, geography, public health, sociology, anthropology, literature, film, and environmental science have incorporated the concept of solastalgia into their work. The concept underpins many new competitively funded and peer-reviewed research programs, especially those that relate to climate change and mental health. Postgraduate students have identified solastalgia as an important new theme in fields as diverse as poetry and fire disasters. Internet-based searches, on sources such as Google Scholar, show that there are now dozens of honors, master's, and doctoral theses written with solastalgia as a major research theme. In addition, some governments, international bodies such as the Intergovernmental Panel on Climate Change, and medical journals with global reach such as *The Lancet* have discussed solastalgia, in their publications, as one of many potential and actual mental health impacts on the mental health aspects of climate change.[31]

The concept of solastalgia has now been applied to the study of negative environmental change in many countries, including Canada, Indonesia, Ghana, Mali, Pakistan, Germany, China, the United States (Alaska, Appalachia, and Louisiana), and the United Kingdom. Common themes are the impacts of mining and climate change on emplaced people and their communities. Solastalgia is now commonly used to evaluate the impacts on people and communities of wildfire, flooding, and other climate change–related events.[32] Other global development pressures in the form of land clearing, overfishing and overhunting, intensive agriculture, and rapid urban change have all been topics of investigation.

In *The Edge of Extinction*, Jules Pretty writes of his travels to many regions where the natural world and people are under increasing pressure from development.[33] As he describes cultures being lost and home environments irreversibly transformed, there is a powerful sense of solastalgia conveyed to the reader. On a trip together to the Karratha region of Western Australia, Jules and I experienced that sense of loss firsthand at Murujuga, on the Burrup Peninsula in North West Western Australia, as we explored the many petroglyphs that can be found there. In addition to their accurate depiction of the animals of that region, including the now-extinct Thylacine or Tasmanian Tiger, these petroglyphs are notable because of their antiquity. Some of the rock art on this site has been dated at thirty thousand years and may well be a lot older.[34] It is known that the original people of the region, the Yaburara, occupied this area until 1868, but after a series of massacres inflicted on them by government-backed forces, they ceased to exist as an integrated tribe occupying their own territory. Jules and I could not help but feel waves of solastalgia as we physically witnessed the fact that this site, so important for Aboriginal people, and indeed all humanity, was now developed as a natural-gas export hub and a location for other petrochemical industries, including an explosives factory.

In Ghana, Petra Tschakert tested the relevance of the concept in communities affected by drought. She found that as people experienced place desolation under climate-enhanced drought, the loss of their endemic sense of place could be described as solastalgia.[35] In Canada and Alaska, the impact of both climate change and development pressure combined to produce a range of significant psychological and emotional responses in Native people, including solastalgia. Ashlee Cunsolo Willox and her colleagues have also used solastalgia to describe the experiences of Inuit communities in northern Canada coping with the effects of rising temperatures.[36] The loss of glaciers in mountainous regions has also been identified with solastalgic feelings. The loss of the Comox Glacier on Vancouver Island in British Columbia has been felt as an emotional loss defined as solastalgia by traditional people in that area.[37] In Alaska, a report by the Alaska Division of Public Health has highlighted the likely impact of unwanted environmental change and the likelihood of solastalgia as a consequence.[38]

On the Torres Strait island of Erub, Karen McNamara and Ross Westoby from James Cook University in Queensland have revealed that respected older women (or aunties) are experiencing solastalgia because of changes to their home environment, including tidal surges, increasing inundation, and altered trends of weather, flora, and fauna.[39] Rising sea levels worldwide are having similar impacts on people in Florida and other low-lying places such as Bangladesh and the Marshall Islands. Kathy Jetnil-Kijiner, a poet from the Marshall Islands, gave an address on September 23, 2014, to the UN Secretary-General's Climate Summit. She performed a poem titled "Dear Matafele Peinem," written for her daughter.[40] The journalist Eric Holthaus wrote, in response to this poem-letter, that it "is an example of solastalgia, the increasingly pervasive feeling of sadness and loss for a world that's being irreversibly altered."[41] I have to agree.

In low lying coastal locations in Louisiana such as New Orleans, the effects of Hurricane Katrina, the BP-operated Deepwater Horizon oil rig explosion and spill in the Gulf of Mexico, plus sea level rise have triggered multiple research efforts partly based around the experience of solastalgia. Indeed, the first international conference on the concept was named "Solastalgia: Longing for Home without Ever Leaving," held March 25–26, 2011, in Lafayette, Louisiana. It was coordinated by the highly skilled folklorist filmmaker Conni Castille. I was pleased to be there as keynote speaker.

Volcanic eruptions, and the aftermath of earthquakes and tsunamis, have been studied as sites of solastalgia after the initial disaster has abated. A number of writers have linked it to the aftermath of earthquakes in the New Zealand city of Christchurch. In places like Indonesia, solastalgia and the environmental distress scale developed by Nick Higginbotham have been used to "examine the psychosocial and environmental distress resulting from the 2010 eruption of the Merapi volcano and explore the experience of living in an environment damaged by a volcanic eruption."[42]

The impacts of Hurricanes Irma and Harvey in the United States in 2017 have also triggered further interest in the relevance of solastalgia to the waves of human-enhanced natural disasters now occurring with increasing frequency and intensity. Worldwide, these events are causing people to become more aware of negative psychoterratic states attached to extreme weather.

In Appalachia, where mountain-top removal for coal has taken place over many decades, a number of scholars, health practitioners, and citizens have, not surprisingly, seen the relevance of solastalgia to their plight. There is both a research effort to document the psychological and physical impacts of coal mining and community activism to prevent further ecosystem and human health damage.[43] Maria Gunnoe, from Appalachia, who received the Goldman Environmental Prize in 2009, fought against the causes of solastalgia in her personal life and community. I was able to get a glimpse into Gunnoe's mind when she wrote about her experience in 2009:

I'm settin' there on my porch, which is my favorite place in the whole world, by the way—I'd rather be on my front porch than any other place in the world and I've been to a lot of places. As it stands right now, with the new permits I saw last week, they're gonna blast off the mountain I look at when I look off my front porch. And I get to set and watch that happen, and I'm not supposed to react. Don't react, just set there and take it. They're gonna blast away my horizon, and I'm expected to say, It's OK. It's for the good of all.

> Am I willing to sacrifice myself and my kids, and my family and my
> health and my home for everybody else? No—I don't owe nobody nothin'.
> It's all I can do to take care of my family and my place. It was all I could do
> before I started fightin' mountaintop removal. Now that I'm fightin' moun-
> taintop removal, it makes it nearly impossible. But at the same time, my life
> is on the line. My kids' lives are on the line. You don't give up on that and
> walk away. You don't throw up your hands and say, Oh, it's OK, you feed me
> three million tons of blasting material a day. That's fine, I don't mind. It's
> for the betterment of all.[44]

Finally, in the context of Australia and my much beloved South West
region, the academic John Ryan, in his wonderful book *Green Sense,* has used
the term "solastalgia" to describe the loss of botanical memories and the
degradation of once botanically rich landscapes in South West Western Australia.[45]

Solastalgia in Popular Writing and Culture

Internationally renowned authors, such as Naomi Klein in *This Changes Everything,* have given significant prominence to the concept of solastalgia, and it is
now well established in mainstream literature on the interrelated climate and
environmental crises.[46] Solastalgia is also referred to by Richard Louv, who
wrote the influential and best-selling book *Last Child in the Woods*, in which
he coined the term "nature-deficit disorder."[47] This new, nonmedical, term
describes the possible negative consequences to individual health and the social fabric as children move indoors and away from physical contact with the
natural world. The book sparked a national debate in the United States and
inspired an international movement to reconnect children with nature. In the
follow-up book, *The Nature Principle*, Louv examines the similarities and differences between nature-deficit disorder and solastalgia, and lays out a systematic framework for overcoming nature-deficit disorder.[48] As an environmental
educator for much of my working life, I have found his work inspirational.

In blogs and other forms of personal writing, solastalgia is now connected to a huge list of issues and events worldwide, where authors have
seen the connection between their emotional states, with respect to environmental themes, and my concept. That human-enhanced natural disasters
are increasing, and that insults to the Earth are becoming larger and more
frequent under global development models, means that more people are

now directly experiencing negatively perceived environmental change. Unfortunately, this means that reports of solastalgia are on the increase. Fortunately, caring people such as Trebbe Johnson, in her 2018 book *Radical Joy for Hard Times*, have used solastalgia as the starting point for acts of repair for damaged places on Earth. I have been pleased to participate many times in Trebbe's annual "Earth Exchanges," where a symbolic act of healing a damaged site takes place.

In the creative arts world, I have seen solastalgia grow as a source of inspiration for a wonderful range of artists and creative people. Writers and artists have always intuited solastalgia in varying degrees, since the underlying negative psychoterratic state is not new to humanity. Edvard Munch's famous painting *The Scream* was indirectly a solastalgic personal response to global environmental impacts, as a result of the eruption of the volcano Krakatoa in Indonesia in 1883. The graphically depicted blood-red sky was a by-product of volcanic dust ejected into the global atmosphere. I once wrote that "Munch produced an archetypal, eco-apocalyptic response in the famous painting. In his journal of 1892, he wrote that he felt as if '. . . a great, infinite, scream [had passed] through nature.' "[49]

A similar degree of existential distress over environmental disturbance can also be found in the work of surrealist artists, such as Salvador Dalí's response to the desolation of mind and landscape as a consequence of war and other transformative powers. Romantic and nature poets have also contributed to the theme of the gradual loss of a loved home environment.

Art and artists have a hugely important role in helping others understand what is going on in their surroundings. Art and craft help us see and react to what is often almost invisible and unspeakable. They bring it all to the surface and force us to interact with it. Artists can offer us a degree of solace with their creations, but they can also heighten our psychic discomfort in the confrontation with environmental disturbance. Contemporary environmental art portrays the loss of species and ecosystems as something more than a loss of biodiversity. It also depicts the loss of something vital within us, the negation of the very possibility of deriving happiness from our relationship to the environment. That can be very unnerving.

Artists not only sense the alienation that is occurring to human-place relationships, they also depict such relationships in their art. When presented with the conceptual clarification of their inner feelings about Earth relations, they are empowered by it. Informed by their knowledge of solastalgia, dozens of contemporary artists, the world over, have created both individual works and whole exhibitions based on the concept.

In 2012, I participated in a traveling exhibition curated by the Lake Macquarie Art Gallery of New South Wales called *Life in Your Hands: Art from Solastalgia* and was involved in the creation of one of the exhibited works. In 2008, with Allan Chawner, a renowned Australian photographer and a former colleague of mine at the University of Newcastle, I flew in a helicopter over the mined areas of the Upper Hunter to film and record images of what people usually perceive only at ground level. The open-cast coal mines and the coal infrastructure were, from the air, re-vealed for all to see. From that collaborative excursion, Allan Chawner produced a strong photography-video piece for the exhibition that tied solastalgia to the utter desolation of the Upper Hunter Valley. Robyn Daw, the exhibition curator, wrote an essay for the educational support mate-rial for the show:

> *Life in Your Hands* maintains an optimistic position in that the potentially redemptive nature of the act of making art can offer some understanding of the notion of solastalgia, provide a platform for discussion and offer a creative response. The search for resolution of issues is ongoing, and art can play a significant part in the restitution of a sense of well-being in a community.[50]

In addition to the many artworks and exhibitions, there has been an ex-plosion in other types of cultural productions creatively inspired by solas-talgia. Plays, poetry, dance, song, and musical composition have all flowed as creative responses. In late 2017, an Estonian composer, Erkki-Sven Tüür, wrote a concerto named *Solastalgia* for piccolo which had its world premiere on December 6, 2017, with the Royal Concertgebouw Orchestra in Amster-dam. Tüür, who lives on the island of Hiiumaa in the Baltic Sea, explains his reason for naming the composition:

> I notice how my life in this amazing place is becoming more and more illusory. . . . It is here that I experience the effects of global climate change first-hand. There's nowhere left to hide from the changes caused by human activity. When I see the massive deforestation here on Hii-umaa or the rampant spread of palm oil plantations when I travel to South-East Asia . . . , I'm overcome by a terrible, sinking feeling. . . . What I actually want to say is that the title of this work isn't just some fashionable term that has a nice ring to it—it's my daily reality. It really does affect me.[51]

In 2018, the well-known Australian singer Missy Higgins named her album *Solastalgia,* explicitly connecting the theme of the album to her concerns about the state of the environment and the threats to future generations. Solastalgia has now well and truly entered popular culture.

Animal Solastalgia?

While solastalgia occurs with the degradation of our place and sense of place, Marc Bekoff argues in *The Emotional Lives of Animals* that since nonhuman lives are part of our community landscape, we can "experience solastalgia when we erode our relationships with other beings."[52]

As humans in the early twenty-first century are beginning, in some meaningful sense, to see other animals, indeed other beings, as kin to us, it becomes more and more plausible to understand that nonhuman animals would feel emotions similar to those of humans when their home environments are disturbed. There has been very little research on the likely impact of development pressures and climate change on the health and well-being of wild animals. I have argued at conferences that nonhuman beings are likely to be extremely stressed by the imposed changes of trophic mismatches, habitat destruction, and global warming.[53] I see no reason why a sense of disorientation and distress in animals should not be called solastalgia. It is, after all, a form of place-based emotional distress in a sentient creature that shares a common environment with us. In chapter 4 I will take this idea of a shared environment and a shared life much further.

Solastalgia and Ecocriticism

In the emerging domain of ecocriticism, the application of ecological and environmental ideas to the evaluation of literary and other artistic forms has now become almost mainstream.[54] The role of solastalgia as an environmentally related concept also enabling eco-psychological evaluation, and as an emotional critique of environmental despotism within film and literature (including nature writing), is now proceeding apace.

The late professor Don Fredericksen, by bringing solastalgia and Jungian theory together, produced a powerful way of understanding our current dilemmas. Film can bring distressing events that were formerly hidden from view into the public eye. Coal mining and fracking are now the subjects of documentaries. Those viewing documentary "environmental" films such as *Gasland* and *The Last Mountain* must confront not only the impact of environmental desolation on other persons, but also their own role in the causes of the desolation such as open-cast mining or mountaintop removal of coal. Fredericksen argues:

> A film like *The Last Mountain*, or Albrecht's work in Australian coal-mining regions, brings those lives within our sight. When we subsequently evade acknowledging what we now know, we are coming into proximity with our capacity for evil. I read Jung's *Memories, Dreams, Reflections* to be saying that we need myth that keeps our feet to the fire of our awareness of this very human act of evasion.
>
> This is an exceedingly uncomfortable corner into which Jung backs us. Here I do not want to know what I need to know—indeed, what I do know. Psychologists have long characterized this as "cognitive dissonance." The fact that I share this evasion with others is no solace. I also do not want to acknowledge that lists other than the one for coal give me what I want for sustenance and comfort; the list of lists is too long. But knowing that I know what I do not want to acknowledge yields perhaps the most painful awareness. I mean simply this: by evading my own titanic psychological register, evading it perhaps titanically, I am polluting my own psychic home. This constitutes an inner, self-generated form of solastalgia, a home warring against itself, no longer providing the solace that comes from ignorance. Any notion of a quick solution to this condition is itself titanic. And hope for the working of the transcendental function, for some mediating third, for a symbol of transformation, may be misplaced. But Jung is undoubtedly right in stating that the issue at hand this time is the very "embodied" soul of humankind.[55]

Fredericksen thus takes us back, via Jung, to Freud and Heidegger, and the affinities between the seats of the uncanny, consciousness, the unconscious, and the biophysical "home." The identity of the outer and the inner, the convergence of biophysical and mental desolation, in the form of a mental "home," also remind me of Bateson and Mitchell and their eco-mental landscapes and psychic stability. Solastalgia in ecocriticism brings the human element of environmental desolation to the forefront. It is my hope that solastalgia is a powerful enough concept to carry such a heavy load.

Research on Solastalgia in the Hunter Valley

After creating the concept of solastalgia and witnessing its application world-wide, I find it instructive to look at how my colleagues and I further developed the concept in the context of its place of origin, the Hunter Valley of NSW in Australia. My publications and joint publications have been informed by collaborative field-based research that has sought to find from those emplaced in the region and impacted by a major environmental stressor, open-cut coal mining, just how they felt about this negative change to their conditions of life.

The people of the Hunter Valley who contacted me about their plight were distressed about what was happening to their home environment. "Distress" is a word that can be used in any number of contexts. I needed to find out from these Hunter Valley people whether there was a component of their distress that was connected to solastalgia, defined as negatively perceived environmental change. If I could tease out this component from the "universe" of distress, then solastalgia would have found justification and verification in the lived experience of people affected by the transformational changes taking place in the valley.

Fortunately for me, two of my closest friends and colleagues during my time at the University of Newcastle were a social psychologist, Nick Higginbotham, and an anthropologist, Linda Connor. Both were keen to enter into a research program that would focus on the people of the Upper Hunter Valley and residents' account of environmental change and its impact on their sense of place.

We initially conducted a small-scale research effort based on face-to-face interviews with both community folk and leaders. Our questions were very much targeted at the person-place relationship, and our interviewees were those who responded to our public notices in the region that asked for volunteers willing to be involved in our research. What we found confirmed my own experiences talking to people over the phone about the impact of mining on their lives. We published this work in a series of articles that included both qualitative and quantitative data on the question of environmentally induced distress in people.[56]

A female rancher summed up what, for me, were the distinguishing features of what I had once vaguely thought about as place-based distress. In her interview with the research team, she described the intense psychological and physical pain she and her farm manager felt about the impacts of coal mining on her rural property:

> Well, I noticed when this business with [mine name], when I was really fighting here. And my manager would come to me and say he didn't sleep

last night. The noise, because they're loading right near the road, he's just across the creek from the road. And you hear a drag line swinging around and dumping rocks into a truck. And then the truck would back away . . . beep, beep, beep, beep, beep. And then the next one would roar in. He used to say to me, "We just can't cope any longer." . . . I lost a lot of weight. I'd wake up in the middle of the night with my stomach like that [note: clenched fist], and think, what am I going to do? We're losing money, they won't listen to me, what do I do? Do I go broke? I can't sell to anybody, nobody wants to buy it because it's right next to the mine. What do I do? And I was a real mess.

The "mess" described so graphically here contains the psychological and emotional impact on this woman, but it also indicates that the body is not immune to insults from noise and sleep disturbance. We are complex organisms where all aspects of our being (physical and mental) relate to our overall health.

Another resident of the Upper Hunter Valley reported in an interview that she found the destruction of her home landscape profoundly distressing:

Originally they [the miners] said they were going to go underground but the DA [development application] . . . is for open cut. . . . Now that is the danger. Species there, there is a very rare woodland banksia in all of that. And it's distressing. It almost reduces me to tears to think about it [mining]. When the coal is gone, the people of Singleton will be left with nothing but the final void.

An Indigenous man in the Upper Hunter Valley expressed his disgust at the mined landscape when he was being interviewed. So distressed was he by the sight of the desolation of his homeland that he would drive hundreds of unnecessary kilometers to avoid seeing it.

It is very depressing, it brings you down. . . . Even [other Indigenous] people that don't have the traditional ties to [this] area . . . it still brings them down. It is pathetic just to drive along, they cannot stand that drive. We take different routes to travel down south just so we don't have to see all the holes, all the dirt . . . because it makes you wild.

In addition, many interviewees pointed out that the transformation created by mining not only disturbs the landscape, but also disrupts cultural cohesion and produces anomie. As family farms were purchased by the mining companies, neighbors, and sometimes family members, who wished to stay, were placed under huge stress. There is no legal obligation to sell, but as they said, "Who wants to live next to a coal mine?" Property values plummet as the mine expands, so those who hold out the longest have the most to lose.

The demographic profile of communities changed with an increase in young, single, male mining workers. The high incomes of mining workers and allied industries drove up prices in the shops and the cost of services, and the original nonmining people felt like second class citizens within their own town. As noted earlier, this inflation created tension in the towns, with the "golden handcuffs" preventing people from getting out of high-income mining once they enter because of high levels of indebtedness. The net result of the socioeconomic changes included a higher level of rental properties in areas planned to be mined in the future and degraded farms that were no longer run as viable businesses because their owners worked in the mines. These negative changes to the built landscape were also a source of solastalgic distress for those who highly valued the former rural character of the place.

The impact of open-cast coal mining on whole communities was also evident, with interviewees constantly pointing out that, although the pressure to sell was placed on individuals, the cumulative effect of pressure on many individuals changed the structure of whole areas. One woman expressed the tension in this way:

> It's a very stressful position to be in because these companies really put a lot of pressure on you and particularly for anyone who's been bought out by a mine. Like you're totally displaced and you're basically given the idea that you're going to be put out of your home whether you like it or not, you know? And I think there's a lot of people in the area who have been affected in that way, that have been basically pushed around by heavy industry.

For these people, "home" formerly consisted of a rich mixture of ecological, farming, cultural, and built heritage, with perhaps old-fashioned rural notions of community holding it all together. While the larger towns of Singleton and Muswellbrook had less social cohesion than the small close-knit villages, all who felt that the impact of mining was negative also felt that their whole way of life was being negatively transformed. In the case of people in smaller villages, their homes were subject to such invasive forces that people could no longer live in these locations. Small but once vibrant places were being depopulated, and in some cases the village became deserted, then demolished, as the mine took the land on which the village once stood.

While this process was taking place, the remaining citizens had to live with massive impacts from the dust, noise, blasting, and terraforming of the encroaching mines. As I explained earlier, the respondents in our study remained in an area that was their location of choice or part of their family heritage. But now they experienced the destruction of nearly all aspects of life that once provided them with a sense of place, identity, and solace, tied to the distinctive qualities and features of life in rural New South Wales. All of the essential elements that make rural life highly attractive—clean air; fresh water; clear, silent, starry nights; scenic landscapes; and endemic biodiversity—were being negatively affected by the transformations and pollution inflicted by the coal-based industries. As people experience the diminution of the quality of those elements in their lives, they experience solastalgia. Because mining is long term and relentless, the solastalgia occurs twenty-four hours a day, seven days a week, all year and for decades.

In some respects, solastalgia is also transgenerational, as some of the pioneering colonist families came to Australia to escape the worst effects of the industrial revolution in Great Britain. One male interviewee said:

> One of the reasons they [his ancestors] left the north of England was on the physician's recommendation because they were suffering from respiratory problems and consumption. . . . The child mortality rate was pretty high. . . . They had steam engines roaring past the house and black smoke and soot. Yes, it's gone round in a big circle. It took a hundred and fifty years, they came here to get away from it, and they did. They said what a wonderful country it is and it's caught up, the industrial revolution's caught us again, we've got the same trouble. Where do we go? Patagonia or somewhere?

The qualitative research we undertook in the Upper Hunter region of NSW concluded that solastalgia is a useful conceptual and descriptive term for the combined environmentally induced desolation and powerlessness that impacts people in the zone of affectation of coal mines and power stations. The team continued to work on the impacts and interpretation of major development in the Hunter Region, including new mine proposals and climate change.[57] In the publication, based primarily on the work of

Nick Higginbotham and his environmental distress scale (EDS), we also pioneered an approach to measuring the amount of distress that can be attributed to negative environmental change within an environment that is being transformed.[58] As a transdisciplinary philosopher, I have no expertise in deriving statistical satisfaction from a good Cronbach's alpha, but Nick and I have always agreed that if I could think about solastalgia, and if it was evident in the lives of Upper Hunter people, he could measure its expression as a form of environmental distress. As stated above, Nick's pioneering work in this field has now been applied by many others all over the world, in diverse fields where the impacts of negative environmental change need to be measured.

Solastalgia and Drought

One further area where my involvement in collaborative research has revealed more about the experience of solastalgia has been the focus on the impacts of drought on rural people in Australia.[59] Again, using community-based interviews, the collaborative research team was able to appreciate the similarities between the chronic stressors of mining and drought. In the case of drought, we received testimony from farmers about how both social and environmental factors were implicated in their distress. My attention was drawn to the emotional burden on men and women as their farms desiccated, crops failed, animals perished, and the landscape became barren. In one particularly poignant interview, a female farmer lamented:

> Well I guess we're coming into our fifth year of the drought. . . . Our gardens have had to die because our house dam has been dry. . . . So it's very depressing for a woman because a garden is an oasis, . . . out here with this dust . . . that's all gone, so you've got dust at your back door.[60]

While Australia is famously a land of "droughts and flooding rains," the length and severity of drought under the influence of new factors such as anthropogenic climate change now have profound psychoterratic implications.[61] New research by my former doctoral student, Neville Ellis, into the influence of climate change on farmers in the semi-arid wheat belt of Western Australia, has taken the solastalgia work further into the context of rural climate change.[62]

A Case Study of Bulga Village and the Mount Thorley-Warkworth-Bulga Mine

In addition to conducting regional-scale research on the people of the Upper Hunter, I have been closely involved with the citizens of the villages of Bulga and Milbrodale, also in the Upper Hunter region. They have been engaged in a long-standing battle with the state government of NSW, its agencies such as the Planning and Assessment Commission (PAC), and the mining multinational Rio Tinto. Rio Tinto was the owner of the Warkworth-Mount Thorley open-cut coal mine and wanted to expand the mine with an application to the government for approval in 2012.

The mine sits on the edge of what I call "the coal rectangle" in the Upper Hunter, bounded roughly by the towns of Muswellbrook, Singleton, Bulga, and Jerry's Plains, and covers about 390 square kilometers of river flats and the Hunter Valley floor. The Warkworth-Mount Thorley-Bulga mines exist within a strip of land 16.5 kilometers long, and up to 4 kilometers wide, from Jerry's Plains Road in the north to near the village of Broke in the south. The Bulga mine complex is contiguous with the Mount Thorley mine. The village of Bulga, at the location of the now defunct Cockfighter Tavern, is approximately 4.2 kilometers from the edge of the 2017 boundary of the Warkworth mine. As the mine expansion proceeds, the village of Bulga will be as close as 2.4 kilometers from the development consent boundary for the mine. Bulga is affected by one of the largest coal mining hotspots in the southern hemisphere.

Given that the locality of Bulga is in close proximity to existing open-cast coal mines, it did not surprise me that residents of the region expressed serious reservations about the personal, social, and environmental impacts of the proposed mine extension. Their concerns about the environmental impacts of the existing mine were clearly expressed, as were their emotional and psychological reactions to the combined impact of social and environmental degradation threatened by its expansion. From one of the citizens objecting to the mine expansion in an affidavit to the NSW Land and Environment Court (NSW L&EC) there was clear expression of outrage about what was happening to the environment at a landscape scale in the Bulga region. The citizen observed that

> the landscape between Bulga and Singleton is already hideous and will never be properly remediated. Now they want to destroy the Warkworth Sands

Woodlands—an irreplaceable endangered ecology—and other areas designated non-disturbance areas *IN PERPETUITY*. We have already lost far too much of the environment for the sake of profits for overseas companies. It is time to value the environment over royalties.[63]

The citizen's use of an expression such as "hideous" reveals a deep, emotionally charged reaction to the existing mining-induced desolation being conducted on this part of the Hunter Valley. Another objector argued that "there are approximately 300 people living in the Bulga area, and their lifestyle will be unbearable if the Open Cut extension goes ahead."[64]

The chronic change to the landscape produced by the mine removes the very thing that makes rural life so attractive to many people. Those in the "zone of affectation" are chronically stressed by this loss. The submission of one resident reported:

[Since] about 12 months ago I have been able to see mining vehicles appearing on the spoil heaps at Mount Thorley Mine. From on or about December 2011, I have been able to see Warkworth mining vehicles above Saddle Ridge during the day and at night. Seeing the spoil heaps during the day and the mining lights in the evening is a constant reminder of the mines, and has taken away from the country rural night time aspect of Bulga. The once scenic views I enjoyed from my property have recently given way to ugly spoil heaps.[65]

Other citizens lodging formal objections referred to the sight of the spoil heaps and the "intolerable" mine noises as "a constant reminder" of the hugely negative changes that are occurring to their lives and lifestyle. I saw such testimony as evidence that these people were suffering, among other distress and stress-related issues, environmental stress best understood as solastalgia. The total impact on some people was desolating, and solastalgia was implicated in the loss of a former positive sense of place, as objectors within the "zone of affectation" experienced their loved place being transformed from forested and rural to a mining and mined landscape.

Another group of Bulga citizens volunteered their views on the impact of mining to me in my capacity as an expert witness for the NSW Environmental Defenders Office, which was formally representing the Bulga-Milbrodale Progress Association in the NSW L&EC in 2013. Given that the objector submissions focused mainly on the key environmental issues identified in a formal environmental impact assessment, it was instructive to ask Bulga residents specifically about their relationship to place and the likely impacts of an expanded open-cast mine at Bulga. Those who volunteered were asked a set of questions about the impact of large-scale environmental change in the immediate area. It was clear, from the seventeen individual responses received, that these Bulga respondents feared the threat of adverse impacts on their quality of life and their physical and mental well-being. Their answers indicated a profound sense of desolation about what is happening to their beloved home environment. Even though this was not a formal representative survey, I concluded that such responses were indicative of solastalgia as I had defined it. In particular, solastalgia was directly connected to a loss of a sense of place.

In an echo of the very first solastalgia research, residents of the Bulga region expressed their emotional reaction to the negative transformation of "their place" by the existing mining operations. Once again, the harsh reality that mining was destructive to rural living came across as a dominant theme. One respondent argued:

> When we first came here; the silence, at night, was deafening! City dwellers, who came to stay, found it quite "strange" to be in such profound silence and darkness, and we loved it! The summer night sounds of frogs and crickets were wonderfully soothing and daytime noise of birdsong and agricultural machinery likewise. . . . There is no vehicular traffic except for that generated by the handful of houses here. Eighteen months ago all that began to change, where, formerly, we had been protected, by Saddle Ridge, from mine noise and lights, we saw the slag heaps and the attendant noise and lights go higher than Saddle Ridge. Now we have nights often punctuated by loud roaring noise from trucks accelerating to the top of the heap, followed by the metallic "clang" as the load is dumped and the truck descends the heap to be followed by another. We have had to complain about bright lights shining directly toward the front of our house. We constantly (and accurately) monitor noise and make complaint to the mine. . . . Sometimes there is temporary relief, but the same pattern, usually, is repeated on subsequent nights.

Many other citizens were alarmed by the thought of losing the peace and quiet that typified their previous amenity and lifestyle. It was not simply the physical environment; it was also the loss of a community connection. In small rural communities, people rely on each other for many voluntary services and support. For example, the voluntary rural fire and emergency services are only possible when people cooperate for common interests and purposes. As villages and small towns are depopulated, such institutions and their visible structures (e.g., fire stations) become impossible to run and cease to function. The loss of a sense of place as both home and community was a double blow to the psyche. In addition, the fact that both corporate and state government powers were behind the assault on their personal lives made the injustice even more intolerable. All of the essential elements of my definition of solastalgia were present in this case.

One aspect of the emotional response to place degradation that I had not properly appreciated was the anger that people felt. One respondent, frustrated at the situation, exclaimed:

> It is deeply distressing to see Australia's landscape decimated by companies that don't contribute to the overall welfare of citizens. We are angry to reflect that when we are "mined-out" these people will go back to their countries of origin leaving nothing to show but a scarred and ugly landscape. It makes us angry, stressed and disgusted that the governments that supposedly represent the citizens, pay no heed to anything but dollar signs!

Another respondent gave vent to this anger by saying that "the shame that we have is what the miners have done to the land. I am more than angry! I am pissed off at them. The noise that emanates from Warkworth mine disturbs our sleep and the PAC has allowed them to move 2.5 km closer after mining through Saddleback Ridge. Bastards!"

I think that anger provoked by changes made to the home environment deserves a name of its own within the psychoterratic. However, I shall leave that thought for a later chapter. In addition to anger, many were aware of the stress burden they carried as a result of the encroachment of mining. The mine was already causing stress between partners when there was no agreement about what was a tolerable level of stress. Different people, for example, had different thresholds of stress tolerance for noise impacts. Some even were aware that distress in villagers had tipped over into depression. One respondent elaborated:

> The current increase in noise and dust levels has had an impact on my family. We regularly have disagreements because of the differing views between us as to whether we should phone up when the noise is excessive, etc. This causes stress between us and was not what we intended to have

when we retired to the Hunter Valley. It causes my wife particular stress as the increased noise levels appear to have more effect on her than me.

The way in which people show their concern about the current situation is severe anxiety and stress. This is brought about by the noise, blasting and dust levels and the potential loss of values of their land and the possibility of having to move away from the problem. I understand there are people in Bulga who are suffering from depression because of the mine's activities.

Finally, for some in the village, the impact they felt was not only local but regional as well. They can now see "the big picture" and gather information via new technology and software such as Google Earth. The realization that so many are caught up in the *sacrifice zone* that is the contemporary Hunter Valley helps sufferers appreciate that their solastalgia is not unique. It is shared by so many who are undergoing similar assaults on the quality of their lives. One respondent summed it all up:

> I love the Hunter Valley. It is a piece of a paradise on this Earth. It is so distressing to go on Google Earth and see the images of the industrial scars getting bigger and bigger month by month and year by year. What sort of people are we? Can we stop this madness? Or are we waiting for the resource companies to change the direction? This would take a very long time. Big corporations have only one plan: to do more of the same and to grow forever like a cancer!

Like me, with my own reaction to the desolation of the valley based on my appreciation of the historical perspective given to me by John and Elizabeth Gould, residents of Bulga and Milbrodale with a long family connection to the area felt the pain reverberating like a mining overburden explosion through their history. One person summarized it brutally by saying, "If my deceased family could come back and see Bulga and surrounding areas how it is now, they would die all over again."

In addition to the material being presented to the court, it surprised me to see in major national media, not usually noted for their empathy for environmental issues in Australia, coverage of the proposed mine at Bulga dealing explicitly with the emotional aspects of the proposed expansion. A feature article titled "What They're Saying Is That We Aren't Worth Anything" reported on the views of Bulga residents. One resident said, "It will kill the town. . . . It will become unlivable," while others reported that the desolation of the town would have severe emotional consequences. The article ended with a statement from one town resident who planned to stay in Bulga. He was quoted saying, "I'd be in all sorts of trouble with my ancestors if I walked away. . . . They'd haunt me."[66]

The outcome of "solastalgia goes to court" was that the chief justice of the NSW L&EC upheld the objection made by the Bulga Milbrodale Progress Association to the expansion of the mine. In a landmark decision, Justice Preston concluded that, on all counts, physical, environmental/biological, and social, the mine's economic benefits would not be sufficient to outweigh the negative costs. On the social front he concluded:

> In relation to social impacts, I find that the Project's impacts in terms of noise, dust and visual impacts and the adverse change in the composition of the community by reason of the acquisition of noise and air quality affected properties, are likely to cause adverse social impacts on individuals and the community of Bulga. The Project's impacts would exacerbate the loss of sense of place, and materially and adversely change the sense of community, of the residents of Bulga and the surrounding countryside.[67]

The people of the affected communities thus won a significant victory in the NSW L&EC. Solastalgia, understood as "the loss of sense of place," had been aired in a court of law and was upheld as a reason not to allow a mine to proceed. The citizens then followed up this victory with another one in the NSW Supreme Court, when the proponent and the state appealed the previous refusal to expand the mine.

The L&EC decision should have been hailed as an ongoing important legal precedent.[68] However, the mining industry and the state government decided to change the law in NSW so that economic considerations would override all social and environmental factors when making significant development decisions. When that law was promulgated, the proponent simply put a slightly modified "new" application in for extension approval, and the planning authorities were obliged, under the new law, to approve it. Bulga and its people lost, and Rio Tinto and the state government won.

Based on my definition of solastalgia, the collaborative research and legal case study material from the Hunter Valley of NSW amply support the idea that the lived experience of negatively perceived environmental change is distressing to those who are subject to it. The emplaced context of solastalgia makes it a unique "environmental" stress among all the other factors that can cause distress in the normal lives of people. It is the desolation of a particular place, the chronic assault on its integrity, that is at the core of solastalgic distress, and it deserves formal recognition in all contexts where people, quite rightly, object to the imposition of such an uninvited stressor.

I have always argued that solastalgia (like nostalgia) is not irreversible, in the sense that it may take only the repair and satisfactory restoration of a place to return solace and comfort to those who seek it. When the mining stops and rehabilitation commences, solastalgia begins to fade, just as it does when rain ends a prolonged drought. There can be a time in the future when the people of the Upper Hunter in general, and Bulga in particular, are once again in love with their home as a place that gives them heart's ease, even though it may be a restored or rehabilitated landscape.

One of my earliest respondents with respect to my understanding of solastalgia was Wendy Bowman, an Upper Hunter woman who has been fighting coal mining in the Upper Hunter for many decades. She even lost a family home to coal mining and had to move as the mine became the neighbor from hell; then her second home was also the subject of a coal mining claim. Now in her eighties, Wendy personified solastalgia as the legitimate experience of emotional pain at the degradation of her loved home and valley. Yet at the same time, she exemplified a burning desire to restore the Upper Hunter as a place that could give generously of solace and beauty, and to protect those valuable places for agriculture, such as the rich alluvial river flats that remain. That she, in the same tradition as Maria Gunnoe, has been awarded an Order of Australia and, in 2017, the Goldman Environmental Prize indicates that the burden of solastalgia can be carried by outstanding people and converted into positive community action.

There are many citizens in the region who have worked hard to save the Hunter Valley from even greater physical and psychological desolation. Local nonfiction authors such as Sharon Munro and Scott Bevan have both included solastalgia as part of their personal responses to the threats delivered by coal mining to the region as a whole, to the Hunter River and its tributaries, and to the climate of the world.[69]

Solastalgia has come a long way in the fifteen years since I first introduced the concept. I was never confident that it would be accepted in the context of human mental health broadly and nonmedically defined. One early critic said the word reminded him of a bad case of sunburn! Despite the critics, and those few who willfully misrepresent the idea, solastalgia has now become part of the international vocabulary and the conceptual frameworks used to describe the impacts of natural disasters, development, and climate change on the emotional and psychic lives of people. It now has a life of its own. In Australia, beyond the Hunter Valley, solastalgia has slowly taken root in academic and popular contexts.

I have given you a small selection of the now thousands of applications of solastalgia worldwide, and I expect there will be many more fields of creative endeavor where solastalgia will have a role to play. As I will explain in the following chapters, from solastalgia I have coined many more neologisms connected to the state of our support or home environment. From a humble start, with just one concept, I now have created many more that I hope will also be seen as needed within the English language (and others) to explain and describe our psychoterratic feelings and emotions.

The concept of the Anthropocene has conveniently given me the ideal context in which to further develop the range of negative Earth emotions like solastalgia.[70] Negative Earth emotions in the Anthropocene is the subject of the next chapter. In chapter 4, I present the idea of the Symbiocene, which assists in the formation and documentation of positive Earth emotions. In doing this, I offer a whole transdisciplinary framework for understanding the relationship between humans, other forms of life, and nature.

The remains of a bird on a paved road in Woodville, New South Wales, Australia.
Photograph by the author.

Chapter 3

The Psychoterratic in the Anthropocene

Negative Earth Emotions

Emotions are primary and primordial forces that motivate us to action. As previously explained, the word "emotion" is derived from the Latin *ēmovēre,* to disturb, at the root of which is *movēre,* to move. Once I had created the concept of solastalgia and watched it gradually take off as a valued contribution to the broadly defined mental health literature, I became aware that it must sit within a much larger range of similar emotions and concepts. What they all had in common was a psychic or emotional state tied to the particular condition of a person's biophysical environment.

Emotional experiences tied to the state of the Earth have long been discussed, depicted, praised, and lamented in art, poetry, and nature literature, yet as with the gap in our language identified by solastalgia, there were very few preexisting precise words in English, other than "nostalgia," that I could identify as belonging to this conceptual, emotional family. I set about to rectify this situation by naming these feelings and emotions. I consider this set of feelings and emotions to sit within the full spectrum of human feelings, but I make no claim to be attempting to describe other types of human emotions.

My insight was that both positive and negative emotional states could be thought of as a family of emotions, with each state able to be defined by a coherent concept. I created the "new" realms of the "psychoterratic" and the "somaterratic" to explain the relationship between the Earth and physical and mental health states.[1] The psychoterratic deals with the health relationship between the psyche and the biophysical environment (*terra* = the Earth), while the somaterratic is focused on the health relationship between the body (*soma* = the body) and the biophysical environment.

Somaterratic Health

Examples of negative somaterratic health issues are the effect of arsenic in the drinking water on people who use wells drilled into arsenic-rich earth and water contaminated by pollutants such that people become ill. The negative impact of diesel pollution and other forms of particulate matter on human respiratory health is another example. The impact of climate change could also be seen as an example of a somaterratic health impact in that, as global temperatures rise and heat waves become more extreme, death from heat stress is likely to become more common. We have already seen this occur over the last four years in places like India, where heat waves have claimed at least 4,620 lives.[2] While I am very interested in these fields of health, I leave this research to medical experts and citizen science activists.

On the positive side of somaterratic health are studies that indicate the health benefits of being immersed in natural settings. The idea of "earthing" (or grounding) in the form of direct skin contact with the ground is an underresearched positive somaterratic state.[3] The works of Jules Pretty and Richard Louv support the health benefits of a "dose of nature" in the prescription for life.[4] Pretty summarizes the importance of this form of support for health:

> At the University of Essex, we have worked for 15 years on how Nature produces mental and physical health benefits. We call this "green exercise." It works for all people, young and old, rich and poor, all cultural groups, in all green environments whether urban park or Nature reserve, whether wild or farmed, small or large. We have shown that a five-minute dose of Nature brings immediate wellbeing. All activities work too, and most people receive an additional benefit from social engagement—doing things together.[5]

I am fully supportive of experiences such as "forest bathing" (*shinrin-yoku*) that give back to the body the elements it needs for physical health—for example, enhanced oxygen levels. I do this at my property at Duns Creek. The subtle combination of trees, running water, wind, and birdsong all combine to give my body what it needs to rest and recuperate its energy. There are, of course, positive mental health benefits from forest bathing, but that will come in the next chapter.

Despite the increasing importance of somaterratic health, it was to the psychoterratic health domain that I felt most capable of contributing something new. Solastalgia was a start, but there was more to do, since, as E. O. Wilson once wrote, people will "travel long distances to stroll along the seashore, for reasons they can't put into words."[6] I felt the need to put into words these reasons and their linguistic emotional correlates, for we are at a time in history when the presence of positive psychoterratic states, in particular, can no longer be taken for granted.

Raw Psychoterratic Emotions

In the past, these Wilsonian positive psychoterratic experiences could be freely undertaken in the world. They were easily had almost anywhere, so we had no real need to put them into words. Now, however, a walk along a beach in a (once) pristine location in a remote area could find a person confronting thousands of bits of plastic washed up on the shore or an oil slick, complete with dead birds and marine animals. As Bill McKibben remarked in *The End of Nature*, the meaning of nature changes when pure feeling no longer governs the experience of wild land or a pristine lake, and incompatible human impacts intrude. He laments the loss of pure feeling when his local lake is "invaded by motorboats":

> A few summer houses cluster at one end but mostly it [the lake] is surrounded by wild state land. During the week we swim across and back, a trip of maybe forty minutes—plenty of time to forget everything but the feel of water around your body and the rippling, muscular joy of a hard kick and the pull of your arms.
>
> But on the weekends, more and more often, someone will bring a boat out for waterskiing, and make a pass up and down the lake. And then the whole experience changes, changes entirely. . . . It's that the motorboat gets into your mind. You're forced to think, not feel—to think of human society and of people. The lake is utterly different now, just as the planet is utterly different now.[7]

The ability to "forget everything" and to live only within the world of feeling and immediate experience is something that humans the world over could once freely experience. For the most part, in benign environments, the forgetting can be a pleasant and stress-free experience. In environments that contain dangerous creatures, one can also freely experience a form of anxiety. However, the fears and anxiety come from within the experience of wild nature and, as part of our evolutionary history, humans have an element of fear as a normal experience of life.

I have fear experiences at Wallaby Farm, as I live alongside two of the most poisonous snakes in the world. The Eastern Brown snake (*Pseudonaja textilis*) is reputed to be the second-most venomous land snake in the world. I live too far away from a hospital and antivenom to survive a bite from this snake. "My" Eastern Brown is about two meters in length, and we have encountered each other a number of times on our property. At close quarters, I have looked into the eyes of this snake and it into mine. We both know that we could kill each other, so there is an uneasy truce. I also know that this species of snake can be aggressive and will attack. It does not know that I am a pacifist and would much prefer no conflict between snake and farmosopher. My anxiety while walking around the farm is heightened on a hot summer's day, when I know "snakey" will be out, warm, hunting, active, and fast. I have surprised it a couple of times, and we both got a shock when that moment to fight or flee was a split-second one. Lucky me.

These are raw emotions, raw feelings of being alive and experiencing life at the sharp edge. Yet although part of our evolutionary background, an everyday fear of snakes is disappearing as dangerous snakes all over the world are being killed or "relocated." People who live in cities and urban zones do not normally encounter any kind of poisonous reptile. They live in a snakeless world and have lost the relevance of having a primordial fear or anxiety of snakeyness. For these people, the loss of snakes is one dimension of the "end of nature."

McKibben's argument that "nature has ended" is also an attempt to get us to see our own culpability in the change from small-scale local and endemic impacts to large-scale and global impacts. Humans have now developed the planet with such force, and with such pervasive technologies, that it is no longer possible to be alone in unmediated nature. The jet contrail overhead, the invisible nuclear fallout, jet skis, helicopter blades noisily chopping the air, and, of course, evidence of human occupation in the form of rubbish—all negate wild nature. Plastic now takes the shape of . . . everything! The human presence begins to overwhelm all else as nature is silenced in the androphony or anthrophony, and the anthropogenic transformation of the whole Earth.[8]

Emotional Death

In this chapter, it is the negative psychoterratic emotions, and feelings about these transformative processes, that are discussed. I write about these negative feelings first, because people can now sense that something is going wrong with our emotional relationship to the Earth. We have gone from worrying and being anxious about intensely local transformation to fears that are now global in their scale. As humans become more powerful as a species, our impacts, especially on ourselves, also become more powerful, so much so that a type of emotional death with respect to nature is taking place.

The emotional death I am thinking about occurs when some humans no longer even have a reaction to the end, death, or loss of nature. There is no emotional presence to bear witness, as all remaining biota are ignored as irrelevant to the life projects of individual humans. With technological isolation from raw nature in the digital age, this form of emotional death becomes commonplace. Distracted by, for example, the small screen, people no longer notice nature. It is no longer physically or conceptually out there: it effectively no longer exists. I had an insight into this form of emotional death while reading J. A. Baker's book *The Peregrine*. In one of the rare passages where Baker acknowledges another human presence in the landscape, he notices something important:

> Fog lifted. The estuary hardened into shape, cut by the east wind. Horizons smarted in the sun. Islands grew upon the water. At three o'clock, a man walked along the sea-wall, flapping with maps. Five thousand dunlin flew low inland, twenty feet above his head. He did not see them. They poured a waterfall of shadow on to his indifferent face. They rained away inland, like a horde of beetles gleamed with gold chitin.[9]

It never ceases to amaze me that, in a place like eastern Australia, there can be untold thousands of migrating Yellow-faced Honeyeaters (small passerines) passing overhead as they move north in the autumn. They are also noisy, with their "ticks" and "tocks," as they chat to each other on their journey. Yet like Baker's man on the seawall, nobody notices them, nobody hears them. People are deaf to the orniphony. It could be the best thing that anyone could see and hear on a bright autumnal day, five thousand happy yellow faces, yet they may as well not exist. Not even a tiny, momentary shadow on the smartphone screen registers as another delicate life in transit. The surrounding birdcall is not even thought of as the ring of another smart phone.

They hear . . . nothing. They see . . . nothing. Their senses and their emotions for connecting to life outside the Anthropos have died.

For those aware of this process occurring, emotional sickness leading inexorably to earthly emotional death is something to be avoided. The negative psychoterratic experiences in life are felt as negative because their "owners" still love life, still want to have life and nature within their own lives. They also show others the flight of the honeyeaters in the hope that environmental education is transformative.

I see the extremes here as three or maybe four different kinds of nature needed to explain where humans are with respect to their experience of nature and life. There is "first nature," where, as for the swimming McKibben, there is a complete merging of the self and the body with the greater forces of the Earth. There is "second nature," where humans are still partially connected to first nature but forge their own technologically mediated Earth. And then there is "third nature" where, whatever nature is, it is no longer normally part of a totally technologically mediated human experience. There is a story of engagement, alienation, and then separation of humans from nature taking place here. There is, of course, a "fourth nature," where reintegration with first nature takes place.

As part of the ongoing separation process, humans experience a number of existential extinction events. We lose culture, language, and our emotional choreography in relation to disappearing first nature. In addition, we experience biological extinction. However, as we lose our emotions and the language that describes them, we lose contact as a species with our evolutionary past. We disconnect from the tree of life and, in doing so, start a process of self-annihilation and incipient insanity.

We can read history to see how this has happened to humans before, as indigenous people were put on this path by colonial powers in places all over the globe. Their experience of emotional loss, then extinction (in some cases), presents a glimpse of our own future.[10] Indigenous literature now contains dystopian plots that tell of solastalgia and other powerful negative displacement emotions for land and places. In Australia, Indigenous writers, such as Alexis Wright in her terraphthoric novel, *The Swan Book*, transmit the very idea of solastalgic loss to the reader.[11] In non-indigenous literature, the resonance of the eco-apocalyptic in *The Road* by Cormac McCarthy conveys a similar experience for those already thinking about the emotions of displacement and separation due to anthropogenic climate change.[12]

A whole new set of emotions, and language suitable to describe them, emerged during the Industrial Revolution. The emotions become focused on notions of growth and progress that privilege human life at the expense of all other life. This attitude is apparent in the rise of an all-consuming materialism and egoism that push away respect, care, and even awareness of other forms of life. The rise of meat eating in affluent classes of humans in the industrialized world exemplifies this emotional disengagement with life.[13] Perhaps even worse is the pushing away of respect, care, and even awareness of other humans who are at the base of the inequality pyramid within the global context. The poverty, dangerous working conditions, and sheer exploitation behind the fashion industry might best illustrate the human dimension of this disengagement with the emotions of gross exploitation.[14]

The mega-theme in cultural evolution, separation from nature, is now encapsulated in the concept of the Anthropocene. Nature and environmental writers such as Robert Macfarlane find it necessary to recatalog the words once used to describe the Earth and its landscapes.[15] Without such conceptual heritage, we are in danger of forgetting that, not only was the world once highly diverse, but that human cultures also had the linguistic nuances to describe it. Language extinction goes hand in hand with endemic landscape and biota extinction. Along with the death of languages that describe the Earth comes the death of the earthly yet largely anonymous emotions that go with them.

The death of Earth emotions is tied to a separation story that has been told by many before me, but I want to reframe it for the twenty-first century. As indicated in the introduction, instead of being a terranascient, or "Earth creating" life form, we have become Earth destroyers, or "terraphthorans." I invented these terms after reading about *Phytophthora cinnamomi*, a soil-borne water mold that produces an infection in plants called "root rot" or "dieback." My much-loved Jarrah forest in Western Australia suffers from this introduced disease, and it is destroying this noble tree's historical range and vitality. The word "phytophthora" literally means "plant destroyer," so I wanted to apply this powerful *phthora* to my work. As I explained in the introduction, the idea that humans can be Earth destroyers, or terraphthorans, was the catalyst for also creating the idea of humans as "terranascient," giving emphasis to the essential creativity of birth.

I will now describe in terraphthoran terms the period in Earth history known as the Anthropocene. Our terraphthoran emotions range from mild forms of forgetfulness, neglect, and impotence to those that are despotic, necrophilic (characterized by love of all things connected to death), and highly destructive. This chapter is about the separation process described above: how it has occurred and what kinds of emotional responses have followed.

The Anthropocene in the Age of Solastalgia

As flagged in the introduction, a meme, or cultural idea, that is in widespread use in recent times is the Anthropocene.[16] One implication of the claim that we are in a new geological period is that humans have now become so powerful a technological species that they dominate and drive all the significant geological, biological, and climatic forces on Earth. Humans also leave indelible physical signs and signals of their global reach, such as nuclear radiation in the soil, plastic in the guts of fish, a high concentration of carbon dioxide in the atmosphere and nitrogen in the soil, and the extinction of species. The human species is leaving a signature on the planet that will be able to be read thousands of years from now. I argue that if these signs are symptomatic of the Anthropocene, then we must exit the Anthropocene as soon as possible.

Clive Hamilton, in his book *Defiant Earth*, argues that we might not have any choice in the future of humanity on this Earth, understood as our home. Within Earth system science, all past conceptions of human-Earth relationships must be overturned. He suggests, with reference to Pope Francis's 2016 encyclical:

> In the Anthropocene, it is no longer tenable to believe that "our common home is like a sister with whom we share our life and a beautiful mother who opens her arms to embrace us." In the Holocene, to view the world, as the encyclical does, as "entrusted to men and women" was a plausible hypothesis. But no more. Now, when Mother Earth opens her arms it is not to embrace but to crush us.[17]

Another possible implication of the Anthropocene is that it is both natural and inevitable that rapacious humans will continue to exploit and destroy their own home, the Earth. Already there are some who contemplate leaving the Earth to travel to new planets or asteroids to exploit as yet untapped wealth.[18] This Earth will be "sacrificed" and the Anthropocene will go cosmic, even universal. I leave it to so-called visionaries like Brian Cox, Elon Musk, and Richard Branson to figure out just what abandoning this Earth means for life in general, not simply human life.[19] The emotions of abandonment have a long history in human affairs, and I have no doubt that this topic will generate considerable debate as to what kind of human could contemplate leaving a dying Earth for greener pastures. I discuss this issue further in chapter 6, titled "Generation Symbiocene."

As argued in the previous chapter, I suggest that one of the defining emotional responses to the chronic desolation of the Earth as home has been captured by the concept of solastalgia.[20] As explained, I established the concept of solastalgia based on the unique experience of people in the Hunter Valley of New South Wales, where open-cast coal mining was desolating hundreds of square kilometers of their valley. All over the world, similar attacks on place and people are occurring at escalating rates. The planetary-scale solastalgic distress now felt by many people is generated by multiple attacks on life and its foundations, and by the sheer size of human population and economic growth conducted within neoliberal notions of progress and development. The largest threat comes from climate change, in particular the relentless warming that is taking place.[21]

Beyond my own publications, the connection between climate change and solastalgia was made by Naomi Klein in *This Changes Everything*. Her work gave international prominence to my concern that global warming and its impacts could produce solastalgia writ large. Klein quoted from my essay, "The Age of Solastalgia," where I describe the global expansion of solastalgia this way: "As bad as local and regional negative transformation is, it is the big picture, the Whole Earth, which is now a home under assault. A feeling of global dread asserts itself as the planet heats and our climate gets more hostile and unpredictable."[22] In my lifetime (sixty-five years), the world and its climate, landscapes, biodiversity, and cultures have all been altered in ways that are hugely negative to life in general, and human life in particular. As Robert Macfarlane said, "Solastalgia speaks of a modern uncanny, in which a familiar place is rendered unrecognizable by climate change or corporate action: the home becomes suddenly unhomely around its inhabitants."[23]

I would go further than Macfarlane and suggest that home is becoming more than unrecognizable: it is for many becoming increasingly hostile. As mentioned above with respect to India, areas of the planet, already very hot, such as the Middle East and parts of Australia, are now experiencing heat waves and maximum temperatures, testing, and at times exceeding, the limit of human endurance. By 2050, according to sober climate-change projections, many of these places will be so hot that "home" will become not just unhomely but uninhabitable. The addition of new, more intense categories—for example, "catastrophic" fire risk and perhaps category 6 hurricanes—indicates that the intensity of weather-related disasters is increasing.

What is this period of human history doing to our mental landscapes as it obliterates and blights biophysical ones? In addition to the concepts of nostalgia and solastalgia, negative psychoterratic states in the existing literature, such as biophobia, ecoparalysis, ecoanxiety, ecocide, and ecophobia will be introduced, as will the emergent role of my new terms such as terrafurie, tierracide, tierratrauma, and meteoranxiety. They will all be discussed as responses to the growing terraphthoran pressures of the Anthropocene, and will set the scene for their opposites; I discuss terranascient psychoterratic states in the next chapter.

Table 1. Origin of negative psychoterratic states

Negative state	*Origin, year first used*
Nostalgia	Hofer, 1688
Necrophilia	Fromm, 1964
Ecocide	Galston, 1970
Ecoanxiety	Leff, 1990
Biophobia	Kellert and Wilson, 1993
Ecophobia	Sobel, 1996
Environmental generational amnesia	Kahn, 1999
Solastalgia	Albrecht, 2003
Global dread	Albrecht, 2003
Nature deficit disorder	Louv, 2005
Ecoparalysis	Rees, 2007
Tierratrauma	Albrecht, 2013
Topoaversion	Albrecht, 2013
Toponesia	Heneghan, 2013
Meteoranxiety	Albrecht, 2014
Tierracide	Albrecht, 2016
Terrafurie	Albrecht, 2017

Negative Psychoterratic States

Negative psychoterratic terms in table 1 are responses to the separation process I outlined above that has been ongoing within the Anthropocene. The terms range from mild to those at the extreme edge of emotional and psychological responses to human desolation and separation from nature. Some elements of the typology appear emotionally neutral, but they still explain how we get into negative eco-emotional states. Others are negative emotions, expressed by people who have not yet given up on their attachment to nature and life. Their responses are expressions of frustration and anger that other humans are embracing a terraphthoran existence. Concepts such as ecocide and tierracide indicate that there will be no "good Anthropocene" in the future, only destruction and possible extinction. I have omitted non-English terms such as *uggianaqtuq* and *koyaanisqqatsi* from the typology. As indicated in chapter 2, however, they are important elements of the cultural response to negative environmental change.

Toponesia

As coined by Liam Heneghan, "toponesia" refers to the process of forgetfulness of precious places that afflicts us as we leave the world of our childhood and enter adult life. Heneghan reminds us that the experience of loss of place occurs from childhood onward, and that in some respects it is inevitable, as places change and we change.[24] As we have seen, toponesia turns into solastalgia as the pace and scale of change increase in one's home environment. However, the loss of special trees, special places within special trees, cubbies, swing trees, and myriad other places that once defined childhood makes toponesia the starting point for all subsequent loss of the topos. Heneghan writes about place and its loss as depicted in the story of Winnie the Pooh:

> But there is another sadness recorded in Christopher Milne's story, a sadness that most of us experience, I expect: the loss of connection with place, especially a natural one, that happens as we grow older. I propose . . . to call this "toponesia" (from the Greek topos, place, and amnesia, loss of memory). Even if the world stood still, we would still spin away from it, dragged into the orbit of our private economies and that series of mischiefs that we call our adult life. These psychological factors associated with Winnie-the-Pooh—its nostalgia, solastalgia and toponesia—combine to make the stories a surprisingly powerful meditation on place, as much as a source of simple pleasure.[25]

Nature-Deficit Disorder

Over and above the normal processes of forgetting the past, becoming fix-ated on the present, and anticipating the future, there are active forces at work that separate children from the natural world into which they were born. Richard Louv has diagnosed one major cause of the separation pro-cess and has created the term "nature-deficit disorder."[26] In *Last Child in the Woods*, Louv warns about the negative impacts and human costs of the withdrawal of our children from nature and natural processes. As he argues, "Our society is teaching young people to avoid direct experience in nature."[27] In his later book, *The Nature Principle*, Louv discusses ways that nature-deficit disorder can be countered.[28] In chapter 6, "Generation Symbiocene," I sug-gest ways that both solastalgia and nature-deficit disorder can be negated.

Louv, Jules Pretty, and many others point out that the epidemics of physi-cal ill health and disability (e.g., obesity) and mental health issues (for ex-ample, attention deficit disorder) in our children are closely related to the disconnection now established between children, their ecosocialization, and their physical activity. Without close physical contact with wild places and wild things, as David Sobel has argued, the socialization and education of children are incomplete.[29]

Children in advanced industrial countries are rapidly losing contact with the world outside of the artificial and the technologically mediated. In countries like Australia, they are born and raised in air-conditioned build-ings and are transported in air-conditioned private cars and public transport. Key locations within the community such as shopping centers and cinemas are also climate-controlled spaces. It is not unusual in summer to have in-door sports venues that are also climate controlled. As Australia gets hot-ter, it is likely that an even greater percentage of time will be spent within climate-controlled indoor spaces. That artificial environment, plus the ubiq-uitous use of entertainment via digital screens within those spaces means that many contemporary children are being exposed to less "raw" nature than ever before in history.

It is little wonder that such children have little or no empathy for wild things, or for nature largely untouched by human action. We cannot ex-pect children to suddenly see or comprehend that the sixth great extinction or climate chaos are threats to their own current and future existence. For them, second nature is gradually phasing into third nature. They will also mature into adults that have phobic fear at the appearance of "unexpected" first nature in the form of a spider in the house, or close urban encounters with untamed "beasts" such as foxes or bears. If the doomsayers are correct, and the eco-apocalypse comes rapidly in an angry dawn, nature's return to everyday life will come as more than just a rude shock: it will be the worst nightmare for those who thought nature was the historic past presented as a documentary on a nature channel.[30] The return of first nature into everyday life as raw and unmediated will generate fear and confusion never before encountered by many generations of humans. Such events are already occurring in places like Florida, where king tides are bringing sea creatures such as octopus into people's garages.[31]

Ecophobia, Biophobia, and Environmental Generational Amnesia

With the nature-deficit socialization of young humans, the artificial becomes all-encompassing. It is no wonder then that fear of nature becomes more systemic. David Sobel and others use the term "ecophobia" to describe the overwhelming fear, at times hatred, of ecology or the biophysical environment.[32] To experience ecophobia is to deny the value of biodiversity, the physicality of the Earth, and the processes that make life possible. The concept of "biophobia," or fear of life, has been developed by Edward O. Wilson.[33] As refined by Wilson and Stephen Kellert, biophobia is a selective response to fear of or aversion to "certain living things and natural situations."[34] Partly genetic and partly social, biophobia could move strongly into the social realm as the Anthropocene expands and dominates. We can then speculate that an irrational fear of life (given that we are a life form) will emerge, as humans become more comfortable with artificial intelligence and robotic companions. Once again, children, young adults, and mature adults who harbor this systemic fear of ecosystems and life will not be able to respond empathetically to its ongoing endangerment and extinction.

As each subsequent generation separates from nature and life, there is a widening gulf that in 1999 Peter Kahn called "environmental generational amnesia."[35] In a later publication, Kahn and his colleagues wrote that "the crux is that, with each ensuing generation, the amount of environmental degradation can increase, but each generation tends to take that degraded condition as the nondegraded condition—that is, as the normal experience . . . [hence] . . . environmental generational amnesia."[36] With such limited experience of first nature to pass onto the next generation, each generation accepts an objectively impoverished nature as the norm. As that process continues through generations, nature ends up simply fading and there is "the extinction of experience."[37]

Perhaps another term is needed to capture the essence of the idea that we cannot forget that which we have never known. Not only are we forgetting something that we once knew, as "amnesia" suggests, but as Kahn argues, we actually know less and less about nature with each generation. "Ecoagnosy," given that "agnosy" is a synonym for ignorance, leads to a real nature deficit.[38] Therefore, our children are also suffering from ecoagnosy, a socially induced form of ecoretrogression.

It is disturbing to think that the current generation may be less ecologically literate, less ecologically attuned, less ecologically aware, and less ecologically emotional than previous generations. As a consequence, they may be unable to respond to the enormous risks posed by ecosystem distress syndromes and climate change. The most dangerous thing about ecoagnosy is that those who suffer from it have no idea that it afflicts them.

Ecoanxiety

In psychology, there has been research on a branch of generalized anxiety disorder (GAD) focused on the environment for many decades. The "environment" in this context has meant anything connected to the suite of settings, external to the sufferer, which might be the cause of anxiety. For example, the schoolyard or workplace might be a factor in GAD and hence be investigated as an offshoot of GAD known as "environmental anxiety." However, the investigation of anxiety more specifically tied to what we would nowadays see as "the environment," connected to nature, has created more nuances within anxiety theory.[39]

The origins of new directions in what is now called ecopsychology do not always come from within academia. The concept of ecoanxiety, for example, emerged from the public domain as people began to explore the specific anxiety or stress connected to the degradation of their home environment. In 1990, the journalist Lisa Leff was the first to use the term "ecoanxiety," in a newspaper article that discussed residents' concerns about the pollution of the Chesapeake Bay area.[40] A "green" issue of a major US newspaper in 2008 featured the theme "ecoanxiety" and included the concept of solastalgia in this emerging new field.[41] From that point onward, "ecoanxiety" has featured in many academic and nonacademic publications worldwide.

In a 2011 publication devoted to the emergent negative psychoterratic impacts of climate change, I identified ecoanxiety as "related to a changing and uncertain environment."[42] With future uncertainty being one of the hallmarks of climate change prediction, a generalized worry about the future is now commonplace. For people such as active climate scientists and those who are fully informed about the science, heightened anxiety is a burden carried on a daily basis as yet more information pours in about negative trends in the biosphere. My 2011 chapter has now been used as a primary reference in a number of academic and government publications to define the field.[43] In addition, in a landmark article in 2011, Thomas Doherty and Susan Clayton summarized the literature and clearly identified the movement from popular to academic contexts for many psychoterratic conditions, including ecoanxiety.[44]

An extreme departure from generalized ecoanxiety has been defined as "severe and debilitating worry about risks that may be insignificant."[45] However, as Verplanken and Roy have argued, "Even high levels of ecological worrying (habitual worrying) are constructive and adaptive, i.e., are associated with pro-environmental attitudes and actions, and are not related to maladaptive forms of worrying such as pathological expressions of anxiety."[46]

Ecoanxiety occurs in those who still have an element of concern left in them for "the state of the environment." For some, however, the next big distraction can easily take them back into the labyrinth of the Anthropocene, where anxiety is more frequently associated with such things as Internet speed, competitive conspicuous consumption, and toxic work relationships. There is also a mild version of ecoanxiety that is connected with personal failure to conform to modern environmental standards by, for example, using and recycling waste such as plastic bags.

Meteoranxiety

"Meteoranxiety" is a subset of ecoanxiety that I have defined as specifically connected to the vicissitudes of the weather.[47] While a traditional form of anxiety is tied to known meteorological extremes, such as thunderstorm or tornado seasons, humans can now become anxious about the likelihood of severe weather events via technologies such as satellites that deliver data and forecasts to 24-7 weather channels and in-person to cell phones. In an era of climate-change-enhanced meteorological extremes, this form of ecoanxiety is likely to become more widely felt.

There can also be meteoranxiety about not getting rainfall in a particular place during a dry period or drought, while all around rain is falling, or unwelcome rain heading in the direction of a farm when harvest is about to occur. Satellite imagery of local weather viewed in real time makes such forms of meteoranxiety a real possibility. Climate change is now delivering extreme weather worldwide, and a reasonable response to this heightened risk of weather-related catastrophe is heightened anxiety. Those who live in flood-prone areas, or close to the sea on cliff tops, now have meteoranxiety as soon as severe storm warnings are given by meteorological agencies. A strong wind on a hot day stimulates fire anxiety and sends anxiety about the weather to abnormal heights. People living in high-risk zones become glued to the weather channel screen or smartphone, and the repetition of the same forecast and warnings only increases their anxiety.

I experienced meteoranxiety about the extreme heat in eastern Australia during the summer of 2016–17. On a 47 degree Celsius (116.6 Fahrenheit) day, I felt deep anxiety about the real possibility of an explosive fire in the local eucalyptus forest because of the volatile haze that was in the air. It was an immediate and disturbing feeling, similar to what I experience in a thunderstorm, when lightning is crackling in the air all around me. How close will it get? The summer of 2018 featured a record-breaking heat wave that lasted three months with so little rain that I was forced to purchase water and have it trucked in. Wallaby Farm lost many trees as the drought and heat killed even the hardy native vegetation. I now have meteoranxiety for the whole summer season as it no longer fits within known former extremes.

Mermerosity

The etymological origins of the word "mourning" come from the Greek word *mermeros*, related to "causing worry," and its meaning is associated with being troubled and grieving. I think we should expand the psychoterratic typology beyond ecoanxiety to include mermerosity, which I define as a chronic state of being worried or anxious about the possible passing of the familiar, and its replacement by that which does not sit comfortably within one's sense of place.[48] A form of anticipatory mourning or grieving takes hold of the psyche, and we are unsettled by it. In many respects, this concept has affinities with the idea of "environmental mourning" pioneered by Renee Lertzman.[49]

In the grip of mermerosity, I begin the process of mourning for that which I know will become desolated, endangered, or extinct long before these events unfold. Mermerosity occurs within people who are in regular receipt of information about the extent of climate change and its projected impacts. I often have an anxious feeling about the future that is not as powerful as global dread (below), but nevertheless matches the accumulation of negative chronic change occurring in this world. I am sure that climate scientists and climate policy experts have similar worries. Bill McKibben comes close to explaining the feeling of mermerosity as he laments:

> The end of nature probably also makes us reluctant to attach ourselves to its remnants, for the same reason that we usually don't choose friends from among the terminally ill. I love the mountain outside my back door. . . . But I know that some part of me resists getting to know it better—for fear, weak-kneed as it sounds, of getting hurt. If I knew as well as a forester what sick trees looked like, I fear I would see them everywhere. I find now that I like the woods best in winter, when it is harder to tell what might be dying. The winter woods might be perfectly healthy come spring, just as the sick friend, when she's sleeping peacefully, might wake up without the wheeze in her lungs.[50]

Topoaversion

Topoaversion is a feeling that you do not wish to return to a place that you once loved and enjoyed when you know that it has been irrevocably changed for the worse.[51] It is not topophobia, where you have fear of a place that might prevent you from entering it; topoaversion is a strong enough feeling to keep you from ever returning to visit the place that was once beloved. The concept has its origins in *topos* (place), and *aversion* (to turn away).

Examples of topoaversion occur when people know that special places they visited—say, as tourists in the 1970s—have been transformed for the worse by development. The island of Bali is now off-limits to many who remember its naive and pristine beauty of the past. I had this feeling in 2017, when visiting Stonehenge in the UK for the first time since 1974, when I had been a long-haired, backpacking hippie. What was once a wild, uncontrolled, and semi-remote place, filled with the mystery and beauty of the henge and its environs, is now a World Heritage site with major road access, internal bus transport, a big visitors center, interpretive displays, and strictly controlled pedestrian zones. I don't think I will ever go back there again, as I feel the development has ruined the place for me. That judgment might be unfair, but my topoaversion has set in and is strong enough to prevent a third visit.

As the pace of development quickens, topoaversion is likely to increase in many as a felt emotional response to the changes that take place. It is somewhat ironic, as special places on Earth become major tourism and ecotourism destinations, that management of the impacts of increased visitation demands changes to the way people and their needs are managed. The whole world has now become a bit like Mount Everest, where the actions of so many climbers and their support systems have turned what once must have been the ultimate wilderness experience (alone at the summit) into the ascent of a huge garbage zone, complete with the frozen dead bodies of past unsuccessful climbers. Jamling Tenzing Norgay, son of one of the first on the summit in 1953, has described contemporary Everest as "the world's highest garbage dump," as a result of the litter and abandoned infrastructure. He laments: "These activities have created a major ecological problem. They are also evidence of disrespect by the climbing community, and a disregard for nature by those men and women who believe their personal conquests are more important than preservation of the integrity of a unique environment."[52] Mountain climbers with a sense of respect for nature, the mountain, and its indigenous people, would now have, or should now have, the emotion of topoaversion that keeps them from adding to this already massive problem.

A similar issue exists in Australia where the famous rock in the middle of Australia, Uluru, is a case study in cultural topoaversion. The Indigenous Anangu people of this region do not, for cultural reasons, want people climbing the rock. However, many thousands disregard their explicit wishes and climb in any case. A culturally informed and respectful type of topoaversion keeps many others from climbing Uluru. I will certainly never climb Uluru. Topoaversion could become a more systematic emotion with respect to ecotourism in general when the full ecological and climatic footprint of such a form of travel is calculated. Recent estimates have put international tourism at about 8 percent (and growing) of global carbon emissions.[53] To continue to ignore such information might be a form of escapism discussed below.

Global Dread

Beyond mermerosity is a more serious existential condition focused exclusively on extreme anxiety about the future. Having its origins in a conversation in 2003, "global dread" refers to a psychoterratic condition that anticipates a hugely negative future state of the world.[54] It produces a mixture of terror, extreme anxiety, and deep sadness in the sufferer who exists within such a state. It is a form of hyper-empathy that projects a person in the present into a terrifying vision of an apocalyptic future.

The dread about extreme climate change and its doomsday scenarios is so great that the terror for the future can generate escapist tendencies in some. I have empathized with those who, under the threat of war and violence, have shifted into the emotional space of a form of euphoria. In *The Book of My Lives,* the Croatian writer Aleksandar Hemon describes his experiences in the early days of the war in his homeland. His term "disaster euphoria" could also be psychoterratic:

> Then there was rampant, ecstatic promiscuity. A few exchanged glances, sometimes in the presence of the boyfriend or girlfriend, were sufficient to arrange intercourse. The whole institution of dating seemed indefinitely suspended; it was no longer necessary to go out before hopping into bed. Indeed, there was no need for bed: building hallways, benches in parks, backseats of cars, bathtubs, and floors were just fine. We reveled in *Titanic* sex; there was no need for comfort or time for relationships on the sinking ship. It was a great fucking time, the short era of disaster euphoria, for nothing enhances pleasures and blocks guilt like a looming cataclysm. I'm afraid we are not taking advantage of the great opportunities provided to us by this particular moment in human history.[55]

The futurist Bruce Sterling created the concept of "dark euphoria" in 2009 to describe the zeitgeist of the coming decade, when he suggested that "things are just falling apart, you can't believe the possibilities, it's like anything is possible, but you never realized you're going to have to dread it so much."[56] At that moment in history, for Sterling, the knowledge of a future apocalypse was so overwhelming that he could see a euphoric rush toward multiple catastrophic endpoints. For some, global dread might be exciting; for others, it is just plain terrifying. I come back to Sterling and dark euphoria in the final chapter.

In the face of environmental and climatic cataclysm, the blocking of guilt, and its sublimation in the form of crass consumption and pursuit of pleasure, is one way to counter the huge negative emotional pressure of global dread. Unfortunately, in the sinking ship of climate chaos, disaster euphoria will be short-lived, just as it is in war. The party will come to an abrupt end, the police will arrive, and the colored lights will be turned off.

A process similar to sublimation in response to global dread has occurred in the past with millenarian, or end of the world, movements. There is an urge within many of us to want a form of Armageddon to purge all our failings, and for everything to start all over again in a new context. To understand our predicament, we need to dig deep into our psyches and past ways of dealing with stress and anxiety. Historically, during periods of social upheaval and unrest, people yearned for instant relief and release from that which oppressed them. For example, various religious movements craved an end of time where all evil and oppression would perish in a glorious moment of change. Even a US president once proclaimed that the end of times would deliver us from evils such as nuclear war. The "rapture" at the end of human time would take us into the promised land. Ronald Reagan famously declared in 1983:

> I turn back to your ancient prophets in the Old Testament and the signs foretelling Armageddon, and I find myself wondering if—if we're the generation that's going to see that come about. I don't know if you've noted any of these prophecies lately, but believe me, they certainly describe the times we're going through.[57]

We are going through similar times right now in the early twenty-first century. Not only has the threat of nuclear annihilation returned, but environmental and social collapse seem everywhere imminent and threatening. A dark global dread hangs over the future, taking on many forms that are all already hugely damaging to some people's lives, such as in subarctic regions and sub-Saharan Africa.

The societal aspects of this build-up to Armageddon are not even subtle anymore. Millions worldwide are suffering from depression and other forms of mental illness, new generations are being born with life chances, such as good health, jobs, and home ownership, significantly below those of their parents at a similar age. Material inequality exists now that has never before been seen on Earth, with the top eight wealthiest people (all men) having more wealth than 50 percent of the population of the whole world.[58]

Social dysfunction arising from drug addiction is now rampant where the forces of hopelessness, loneliness, unemployment, homelessness, and powerlessness all combine to create an oppressive social climate. Addiction to highly dangerous substances such as "ice" (crystal methamphetamine) has been seen by many as a contemporary expression of "the opiate of the people."

There is much more: we have created a globally connected world that is destroying us as we are destroying it. Global dread is now interconnected and metastasizing. Many want out and are grasping at anything that makes the journey either smoother or faster. Such escapist thoughts combine to make climate change denial, escapism, avoidance, and anti-environmentalism integral parts of a neomillenarian movement that wants the current state of the world to end.

Some secretly celebrate record temperatures and climate-related disasters. They actually want the world to change fundamentally. It is no wonder that Christian fundamentalists are so prominent in climate denial circles. They have read and understood the blueprint for such an apocalypse. It may not, any longer, be a thousand years in the future, but the thinking is the same. When life becomes intolerable and there seems no way out, prayer and desperate hope for a final end, so that we might start all over, beckon. The nonbelievers in "rapture" religion simply engage in disaster euphoria, take drugs, and drink more.

The response to oppression just might be a desire to see the cause of the oppression come to a grinding or blinding halt. Being involuntarily locked into a complex system that they cannot change forces people into escapism. They will want climate chaos because they want a change that fundamentally alters a state of affairs they crave to be out of. Despite the risks of future disaster with the great warming, in order to banish global dread, any big change will be a good one.

Ecoparalysis

People who are still aware of environmental pressures and want to respond might be caught in another psychoterratic state that is called "ecoparalysis." A number of writers began to use this term around 2007–8 to describe another type of response to ecoanxiety or powerlessness when confronted by potentially damaging geophysical events. I added it to the psychoterratic typology in 2008 after using it earlier in conference presentations and lectures. My own formulation of ecoparalysis was that it was a reasonable response to the dilemma faced by people who could see the enormous scale of the problem confronting the world but could do nothing meaningful at a personal level to solve it. Changing the light bulbs in the house was not going to solve the world's climate change problem. Hence the choice to do nothing was also to remain in a state of ecoparalysis.

A mild form of this type of paralysis also occurs when people can see a technological breakthrough such as battery storage for renewable energy but feel they must wait until it becomes more efficient or cheaper or both before they will commit to purchase it. Of course, that waiting game could go on forever, as the technology is constantly evolving and improving. They want the big battery, but they are waiting for that perfect moment of cost effectiveness, efficiency, and a big sale. They want a bargain that will never arrive. Their fiscal ecoparalysis makes it impossible for them to be a "first mover" and social leader.

According to a psychological way of thinking about ecoparalysis, Lertz-man has argued that "far from being an absence of *pathos*, or feeling, inner feelings of anxiety, fear or powerlessness manifest as a lack of action or a pa-ralysis."[59] Hence, despite what superficially appears as apathy, complacency, disengagement, or an active decision to do nothing, ecoparalysis is a psy-choterratic condition that maintains people in a state of limbo.

Bill Rees gave perhaps the earliest expression of ecoparalysis in a conference presentation where he specifically emphasized its sociological or structural as-pects and the current human dilemma. He argued that "ordinary people hold to the expansionist myth. North American society remains in eco-paralysis. The ecologically necessary is politically infeasible but the politically feasible is ecologically irrelevant."[60] Again, the inability to respond meaningfully to the climatic and ecological challenges that face us is not always an expression of individual apathy or denial: it is an inevitable outcome of the intractable nature of the problems. The fact that they are tied to the very foundations of our pres-ent economy and its allied political system generates dilemmas not seen before in human history. Both voters and politicians are caught in this gridlock.

As we learn more about our ecological impacts and carbon footprint, it seems that every option to retain "life as usual" ends in contradiction. At some point, saving the world by consuming more becomes absurd. Even the idea that we save energy, and hence carbon emissions, by doing business and per-sonal communication electronically runs into the uncomfortable truth that the annual amount of energy required to run the World Wide Web is roughly the equivalent to the annual energy use and carbon emissions of global air traf-fic. Not many people in rich, technologically sophisticated parts of the world are prepared to embrace the full implications of an ecologically protected and severely carbon-constrained world.[61] While many now clearly see the extent and nature of our problems, the threat of hugely negative events, even those that will impact on their own children, seems insufficient to change behavior as usual. I suggest that such gaps between knowledge, values, and behavior are now sources of ecoanxiety and causes of ecoparalysis worldwide.

Tierratrauma

When people possess a deep connection to the Earth, they can experience deep emotional trauma when that connection is directly impacted by pow-erful forces. "Tierratrauma" from (*tierra*, Earth, and "trauma") describes the moment when a person experiences a sudden negative environmental change.[62] That trigger can be delivered either virtually or by direct experi-ence. Now, in addition to the chronic stressor of solastalgia, we have the newly defined acutely felt Earth emotion, the negative psychoterratic state of tierratrauma. It is also worth noting how tierratrauma is differentiated from other psychoterratic conditions. Tierratrauma is not post-traumatic stress disorder (PTSD) because it occurs at the moment the trauma is initiated; it is not solastalgia derived from chronic change, because the change agent is an acute Earth-based existential trauma in the here and now.

Tierratrauma occurs when wildfire is destroying your local area; a patch of remnant bushland is being bulldozed for a new road; you are witness to an oil spill that smothers all life on your beach. These are poignant moments when there is rapid and devastating change to a loved place or important location. My ecoanxiety about the threat of fire can turn in an instant to tierratrauma, as an all-consuming fire front hits my fence line and the house is under ember attack. In 2017, for the first time in California's modern history, a winter wildfire destroyed hundreds of homes and burned huge expanses of land. The toll on livestock and wildlife was also enormous. In the face of a wildfire driven by the Santa Ana winds, wildfire tierratrauma is now a well-known experience in California, even in winter, as people struggle to survive the permanent drought that afflicts this part of the United States.

I also felt a strange kind of tierratrauma when the Fukushima nuclear plant was hit by the tsunami in 2011. While the whole tsunami unfolded as a massive tragedy for humans—indeed all living things in the area—it was at the terrifying moment when the reactor blew up, as I was watching live coverage on television, that I felt powerful trauma radiating through my body, of a Munch-type that I have rarely experienced before. Such is the power of the global media today, it can shift tierratrauma from one side of the world to another in an atomic instant.

Both solastalgia and tierratrauma are certain to proliferate as climate-related disasters increase while the world warms. Fire, drought, storms, and floods are all increasing in intensity and frequency; it is likely that this trend will continue for the foreseeable future. Because of the temporal displacement of the impacts of the pollution humans are emitting today, future generations are guaranteed to experience a greater degree of solastalgia and tierratrauma in their lives than we are in the present day.

Terrafurie

I was asked to create a new word by ecologically minded friends who felt a shared anger about what was happening to the world but could not put that precise form of anger into meaningful English. I responded with "terrafurie," or earth rage.[63]

Terrafurie is the extreme anger unleashed within those who can clearly see the self-destructive tendencies in the current forms of industrial-technological society but feel unable to change the direction of such tierracide and ecocide (see below). The anger is also directed at challenging the status quo in both intellectual and socio-political terms. Terrafurie is anger targeted at those who command the forces of Earth destruction. I think of it as a protective anger, not one that is aggressive. In chapter 6, however, I explore the type of strength and anger needed to protect the protectors of this Earth from harm.

Many before the present generations have felt a precise anger at the destruction of our own support environment and the homes of countless other beings who want nothing other than to live and reproduce. J. A. Baker wrote a short essay, "On the Essex Coast," in 1971, where he argued in support of those opposing the further development of this last remnant patch of wildness in southeast England. While out walking on the coast, he came across a dead oil-soaked diver (Loon). His reflection on how this bird lived and died prompted an outpouring of solastalgia, as well as very precisely directed anger. Baker gives vent to this anger:

> I blunder on across the saltings, in too great a rage to see or hear anything clearly. After a day of peace, I have seen the ineffaceable imprint of man again, have smelt again the insufferable stench of money. A yellow wagtail flits ahead of me, a brilliant torch flaming up into the sun. That at least seems to be still clean, still untainted. Yet who can know what insidious chemical horror may be operating beneath those brilliant feathers?[64]

There is a deep-seated anger here about the injustice of human impacts on the rest of nature, and the death and sickness unleashed by our industrial forces. It is also a political anger, directed at politicians, as he warns the rest of us not to be "soothed away by the lullaby language of indifferent politicians."[65] I can only imagine the terrafurie of indigenous people the world over as they are forced to experience the destruction of their land and culture.

Terrafurie is likely to become a more common emotion as Earth insults become more prevalent and widespread. As in the Upper Hunter of NSW, the technologies now being used to terraform the Earth are so powerful that "rage against the machine," as well as against the owners of the machine, becomes the only sane option.

Terrafurie can also be expressed as a form of environmental rage and creativity. In April 2007, I wrote a terrafuric poem, printed below, that drew together a number of themes. It was written on the release of information about the feminization of the population of Aamjiwnaang First Nation people in Canada and of the wild Snapping Turtles. Twice the number of girls are being born than boys, and the "feminized" male turtles have diminished penis size. Pollution from "Chemical Valley" in Sarnia, Ontario, is thought to be implicated.

Benjamin Chee Chee was a Native American artist, now famous for his portraits of Canada Geese and other birds. He committed suicide in 1977 at the age of thirty-three. I interpret his simple, graceful, and highly structured artworks depicting geese as an attempt to keep the world balanced and in order. As a Native American, he was trying to defeat the personal and cultural pathology and tragedy of nostalgia and solastalgia. The Aamjiwnaang tell of geese trying to land on a pond in Chemical Valley in a cloud of benzene and dying before they hit the water. Chee Chee's suicide, the problems of indigenous people, and the collapse of order in Canada Geese seem connected.

Life Out of Balance

A thousand years of Aamjiwnaang dreams
spirits touched by pure steam
from sweat lodge rocks
that release a culture's memory.

One hundred short years of inversion
in the Chemical Valley
fugitive emissions into every space
Is the maple syrup really sweet?

Geese struggle for formation in miasmic air
benzene tears in the artist's eye
reveal the reasons for
Benjamin Chee Chee's suicide.

Sweet innocent boys go missing
So too the Snapping Turtle penis
shrinking in the chromosome chaos
the Hopi call Koyaanisqatsi.

A hundred long years of restoration
Geese, turtles and children
once again in perfect formation
life in beautiful balance.

Beyond the spontaneous expression of terrafurie, ecoactivism directed at the forces of destruction is a positive outlet for anger. Anger can be transformed into activism. In the past, with the exception of passive-violent actions such as tree spiking to prevent logging (monkey wrenching), ecological activism has been largely nonviolent. As the anger builds and the indifferent or even hostile politicians keep wrecking the Earth, terrafurie just might flip into violence directed at those who allow the "insufferable stench of money" to continue to corrupt and kill life forces.

In chapter 6, I will take this idea further with an examination of the potentially positive side of terrafurie as an expression of ecomasculinity and ecofemininity.

Necrophilia and Eco-necrophilia

At its extreme, the complete lack of empathy for life and hostility toward it can manifest as a form of necrophilia, defined as a love of death.[66] The support for life is overridden by the greed for wealth and productivity beyond what the Earth can supply without the loss of ecosystem vitality. Greed can trump life in the short term. Erich Fromm extended the concept of necrophilia beyond a sexual perversion to a whole approach to life and death. In *The Heart of Man* he argued:

> While life is characterized by growth in a structured, functional manner, the necrophilous person loves all that does not grow, all that is mechanical. The necrophilous person is driven by the desire to transform the organic into the inorganic, to approach life mechanically, as if all living persons were things. . . . Memory, rather than experience—having, rather than being—is what counts. The necrophilous person can relate to an object—a flower or a person—only if he possesses it; hence a threat to his possession is a threat to himself, if he loses possession he loses contact with the world. . . . He loves control, and in the act of controlling he kills life.[67]

The connection between necrophilia understood in this way and the commodification of nature under capitalism is all too easy to make. The tendency to want to put dollar values on nature to value it as property in order to "save" it under a neoliberal model of capitalism is a necrophilous version of turning life and living processes into dead things. Necrophilia becomes eco-necrophilia and, along with the commodification of nature, our Earth emotions become commodified and are ultimately exterminated.

While old forms of eco-necrophilia such as trophy hunting continue unabated, new forms have emerged in the twenty-first century. In the digital world, where death and mayhem can be conducted within realistic games with impunity, a form of indifference to the suffering of others can be transferred easily into a failure even to perceive the suffering of others, including nonhuman beings in the natural world. Such a condition might even be exacerbated by witnessing violence and death in the natural world only via natural history documentaries. The reality of death is never revealed, because our senses have become dull to it, and the stench of death is completely removed from our everyday experience.

Ecocide and Tierracide

"Ecocide," or the killing of ecosystems, has been a term in use since it was created by Arthur Galston in the 1970s.[68] It was provoked by the extensive use of herbicides such as Agent Orange during the Vietnam War. Such was the scale of damage inflicted on the forest cover of Indochina that Galston correctly thought the world needed a new concept for such large-scale anthropogenic desolation of the Earth's ecosystems. The term has had considerable longevity, with people such as Polly Higgins taking the idea into international law and governance contexts.[69] I tend to think about ecocide and the death of ecosystems as existing at a scale smaller than that of the whole Earth. Higgins, in her *Eradicating Ecocide* campaign, has joined the concepts of ecocide and solastalgia in her recent work. She maintains:

> Ecocide adversely impacts on many levels; there can be harm both ecological and cultural. Our emotions and our senses are affected; we feel and see the adverse impact of ecocide. Communities most harmed by ecocide suffer what is known as solastalgia. At a collective level communities feel a profound sense of isolation and intense desolation. This is compounded by the community's lack of power in the face of State and corporate might, the pain of being unable to console in times of great distress and the loss of homeland.[70]

The unthinkable conclusion of the Anthropocene, with its terraphthoran tendencies, is the extinction of all complex life on Earth: tierracide. This is the deliberate desolation of the whole biosphere, such that it can no longer sustain life-supporting processes. Once we get to ecocide (regional) and tierracide (planetary) as possibilities within the Anthropocene, we need revolutionary thinking to get us out of such a fate. In the next chapter I aim to provide just such a revolutionary platform. After all, to avoid ecocide and tierracide, we need to change everything.

There remains the difficult question of whether humans are capable of making the necessary changes and the possibility that we are, in fact, a necrophilous species. Bill Rees has provided his answer to the question:

> I trace this conundrum to humanity's once-adaptive, subconscious, genetic predisposition to expand (shared with all other species), a tendency reinforced by the socially constructed economic narrative of continuous material growth. Unfortunately, these qualities have become maladaptive. The current coevolutionary pathway of the human enterprise and the ecosphere therefore puts civilization at risk—both defective genes and malicious "memes" can be "selected out" by a changing physical environment. To achieve sustainability, the world community must write a new cultural narrative that is explicitly designed for living on a finite planet, a narrative that overrides humanity's outdated innate expansionist tendencies.[71]

In the following chapter, I will question the innateness of the terraphthoric tendency in human nature and construct a new cultural meme that will put humanity back on track to a terranascient future.

Chapter 4

The Psychoterratic in the Symbiocene

Positive Earth Emotions

There was some hope, starting three decades ago, that concepts such as "sustainability," "sustainable development," and "resilience" would provide the conceptual and practical foundation for relationships between humans and nature on the planet.[1] Despite the honest and best efforts of many academics, policy experts, and conservation bodies, all the so-called indicators of sustainability have gone backward. A key indicator, carbon dioxide concentration in the atmosphere, went from 340 parts per million (ppm) in 1985 to over 400 ppm in 2016. For the month of April 2018 the average was 410.26 as measured at the Mauna Loa Observatory in Hawaii.[2] If we accept 350 ppm as a supposedly safe level for carbon dioxide for preventing further extreme warming of the planet, then quite clearly we have failed to be sustainable and to develop sustainably, as a stable and livable climate is the basis for all other human activity.

The ethical principles built into the very foundations of sustainable development—inter- and intragenerational equity, the precautionary principle, and interspecies equity—are all being violated on a grand scale. The warming of the climate represents a massive failure to respect the interests of future generations of all species. The world, as we saw in chapter 3, has become more inequitable over the last fifty years, with concentrations of

Loraine Teasdale Garden Art in Margaret River, Western Australia.

Photograph by the author.

wealth in the hands of fewer than ever. Such intragenerational inequity will make for even greater levels of intergenerational inequity into the future.

Intergenerational inequity includes nonhuman beings, for as climate change becomes more severe, and the risks involved ratchet up, influenced by the combined forces of global-scale development and climate change, the endangerment and extinction of species will go beyond already dire levels.[3] The failure to enact adaptation policies, in the face of the threats we have already imposed on ourselves, has been sadly illustrated by the severe impact of Hurricane Harvey in Houston in October 2017. We seem unable to prepare to adapt to a known change, and as a result have violated the precautionary principle, one of the key ethical responses to what once was called the sustainability challenge.

I had great hope for a sustainable future for humanity under sustainable development, as I was once a professor of sustainability in an Australian university. However, I no longer have optimism for this conceptual framework in the year 2019. Clearly, by themselves, ethical and policy principles based on sustainability are not sufficient to change human motivation and behavior. They do not stir the terranascient emotions and are too easily incorporated by terraphthoric forces. As a change meme, the concept of sustainability has been rendered inert at best and subversive at worst.

The concept of resilience has also been appropriated by forces determined to pull it into the gravitational influence of industrial society on a globalized scale.[4] Instead of helping us rebound into configurations of successful models of living after disturbance, we are now seeing resilience being used to justify the ongoing existence of processes and activities that are driving humans, and the rest of the globe's biota, to extinction. Coal, oil, and gas-fracking industries now use their advertising and public relations to spin the message that their industries are not only sustainable but resilient as well. The ongoing resilience of nonsustainable and undesirable features of social systems can be termed "negative resilience."[5] Perhaps more appropriately, it has also been named "perverse resilience."[6] These forms of resilience occur where pathological social relationships that are oppressive and exploitative of humans, nonhuman species, and ecosystems are rendered resistant to change by economic and political subsidies (donations), bullying, intimidation, and vested interests. There is a form of corruption working here that destroys the foundations of life.

There is no evidence, to my knowledge, other than some growth in the use of renewable energy and battery storage, to show that any of our current actions are leading toward a more resilient planet or human cultural/industrial system. Critics of renewable energy are quick to point out that solar panels, wind turbines, and lithium-ion batteries are produced using fossil fuels, massive pollution, ecosystem disruption, and human exploitation. They are, however, a transition to something far better, and I have invested in solar electricity, battery storage, and solar hot water in a big way at Wallaby Farm. Despite how evacuated solar tubes, solar panels, and lithium-ion batteries are made, there is absolutely no doubt that, once up and running, they use freely available, nonpolluting, safe forms of usable energy. However, if they are merely being used to expand growth within the old parasitic industrial system, the Earth will be still subject to major desolation; it will just take that bit longer than it would have if we continued using fossil fuel as the foundation of that economy.

Final proof of our current policy failure is the persistence of climate change denialism, even its ascendency, under the current (2019) US administration. Denialism is another indicator that sustainability and resilience are failed concepts. They have been dismissed easily by social commentators and critics as a United Nations "authoritarian" attempt to curtail the freedom of people and to force on them a form of totalitarian ecosocialism. In the United States, "resilience" is being used as a term by the Trump administration in preference to "climate change" in all government policy.[7] I hereby announce the death of "resilience" as a useful term in redirecting the future of humanity.

I have argued, long before the present, that sustainable development fails to define what it is about development that is to be sustained, other than the process of development itself.[8] Global-scale development, which is not broadly in harmony with greater life and planetary forces, puts us on the path to tierracide, or the death of life on planet Earth. We must change rapidly from an exploitative, polluting, and parasitical society to its opposite.

The foundation on which we are building right now is seriously flawed, and conducive of nothing but great waves of ennui, grief, dread, solastalgia, mourning, and melancholia. The ascendency of negative Earth emotions in the twenty-first century is an indicator, or symptom, that we have got Earth relationships badly wrong! Somehow, humans, who evolved within the matrix of life, freely enjoying the best emotional experiences the Earth has to offer, have socially evolved out of that matrix into an extremely dark emotional space. Mind you, it is not all humans. It is mostly that half of the human population who now live in cities and their urban surrounds. In the Anthropocene, the majority of humans have separated themselves from the rest of nature and life, from having some ongoing vital connections to nature to having virtually none.

The reasons why the great separation has taken place are complex, with just some of the dominant themes being the God-given human dominion over nature within Christianity; eco-alienation under neoliberalism and capitalism; the emergence of hierarchy in complex societies; imperialism and colonialism; and patriarchal development or male domination over a perceived female nature.

I will not attempt to summarize the literature underpinning these grand themes; it is enormous, and, for many, there is not enough time in life for a catch-up on this reading matter. Instead, I will argue that on the basis of no obvious evidence to the contrary, these dominant ideas have held sway for so long that they have come to produce a worldview that is horribly resilient yet mistaken at its core.

The core beliefs underpinning social and nature separation that I wish to highlight as mistaken include individualism, atomism, reductionism, and autonomy based on science (evolution) and ideology (neoliberalism). These ideas imply that humans are separate from the rest of nature; humans are physically and morally autonomous; matter can be reduced to its smallest parts; competition between individuals (survival of the fittest) rules in both nature and society; and that competition in a free market within an economy is an expression of natural, competitive order.

The UK journalist George Monbiot has graphically summarized the narrative for us:

> Our good nature has been thwarted by several forces, but perhaps the most powerful is the dominant political narrative of our times. We have been induced by politicians, economists and journalists to accept a vicious ideology of extreme competition and individualism that pits us against each other, encourages us to fear and mistrust each other and weakens the social bonds that make our lives worth living. The story of our competitive, self-maximising nature has been told so often and with such persuasive power that we have accepted it as an account of who we really are. It has changed our perception of ourselves. Our perceptions, in turn, change the way we behave.[9]

Individualism, separating humans from nature and each other, seems to have become a dominant narrative in the twentieth and now the twenty-first century. These dominant ideas have been challenged for a very long time by some in the arts, social sciences, and humanities, yet they remain firmly embedded in our systems of commerce, law, politics, and ethics.

For an equally long time, the realm of science, another socially dominant conceptual framework, has not seriously challenged these ideas. The science of ecology definitely did require an integrative mentality, to see how organisms were related to each other within artificially defined boundaries, and a few well-known case studies showed how symbiosis, or coevolved, mutually beneficial relationships between symbiont species, existed. Oxpeckers and buffalo, for example, are symbionts that have a mutually beneficial relationship around the removal of parasites. The problem was that ecology was dismissed as a serious science in reductionist scientific traditions, especially in universities. The real world was too messy for controlled experiments.

There is another element as well: many of the recent champions of symbiosis and holistic views of life have been female and were marginalized in their respective male-dominated fields. Elyne Mitchell (1913–2002) was the first such champion in Australia that I am aware of, and she had a very sophisticated view of the regenerative capacity of soil to sustain life:

> In this conception of living soil, man, animals, vegetables, fungal growth, micro-organisms, bacteria—all that is life—form an integral relationship. This relationship is only properly preserved in the cyclic law of return, by which the fertility of the soil is continually maintained by the return of all wastes—from the vegetation and from the animals and from man who eats both animal and plant. So that which was life is returned to the soil where the micro-organisms and the fungi transform it into the current of life once more.[10]

The biologist Lynn Margulis propounded the symbiotic view of life and contributed to some of its research foundations for well over four decades, until her death in 2011. Earlier, Rachel Carson, while not explicitly writing about symbiosis, was forwarding the idea of the "web of life" or "interconnectedness" in nature. Her famous work, *Silent Spring* (1962), gave specific emphasis to the interrelatedness of life when discussing soil. She wrote: "This soil community, then, consists of a web of interwoven lives, each in some way related to the others—the living creatures depending on the soil, but the soil in turn a vital element of the earth only so long as this community within it flourishes."[11] Carson's attack on the role of poisonous chemicals in the air, water, and food produced a "politics of pollution." The male world order, especially in commerce, was founded on the sanctity of the atoms of life and men's ability to control and own them. Commerce in agriculture was built on the idea of the "autonomous life agent" that could be controlled in factory-like settings. Various killers of life, the "cides," were introduced to control weeds, insects, and fungi. Ideas such as "the web of life," ecology, symbiosis, and all forms of interrelatedness ran counter to reductionism and atomism in patriarchal science. That many early champions of ecological and symbiotic thinking were female only added to the threat to the patriarchy, reductionism, and mechanism that have long ruled in academia, science, commerce, and industry.[12] However, the new challenge to Anthropocene individualism has now come from within science itself with a capital *S*.

Science and the Rise of Symbiosis

Biology (the study of life), as one of the hard sciences, now presents irresistible evidence that the rules of the foundations for life are interconnection, interrelationships, diversity, and cooperation, and that homeostasis between the "diversity of lives" gives the stability that life in general needs to endure. Symbiosis has now emerged as a primary determinant of the conditions of life. Kelly Clancy makes the case for symbiotic coexistence as a foundation for life:

> Long periods of harmonious co-existence may be the evolutionary precursor for true symbiotic relationships. Billions of years ago, another ancient cyanobacteria was engulfed and "domesticated" by an ancestor of plants. It shed most of the genes it needed for an independent existence and became what we now know as the chloroplast. In return for a safe environment, these chloroplasts performed photosynthesis for their hosts, fueling a new form of life that eventually spread over much of the Earth. It's likely this same kind of division of labor was a seed for the development of multicellular organisms. Here, evolution is not a weapons race, but a peace treaty among interdependent nations.[13]

The scientific meaning of symbiosis implies "organisms living together," most often for mutual benefit. As a core aspect of ecological and evolutionary thinking, symbiosis affirms the interconnectedness of life within the variety of all living things. It also implies an overall homeostasis, or balance, of interests, since domination of one part over the rest would lead to functional failure. Symbiosis had its origins with the German botanist Albert Frank in 1877 and his identification of a mutualistic relationship between algae and fungi in lichens. He coined the term "mycorrhiza" to describe the mutualistic interconnections between trees and fungi. Later, in 1879, this idea was further refined as "symbiosis" by the mycologist Heinrich Anton de Barry, defined as "the living together of unlike organisms," as opposed to independent, free-living organisms.[14] The cooperation between radically different types of organisms living in close proximity was, and remains, a radical departure from Darwinian models of evolution based on competition between different organisms in a competitive environment.[15]

At its core, symbiosis counters the idea that evolution is inherently and solely competitive, since life also thrives by symbiogenesis, "the evolution, over time, of new behaviors, physiologies, organs, or organisms that are directly attributable to symbiosis."[16] Evolution is driven by both cooperation and competition, and science is only just beginning to shed new light on the cooperative foundations for life that are provided by an understanding of the actions of symbiosis and symbiogenesis.

While we have known about the relationships between plants, their roots, and their macrofungi companions since the late nineteenth century, detailed knowledge about this type of relationship has grown exponentially in the last fifty years. In Australia, for example, scientists are gradually discovering the sheer size and complexity of that symbiotic integration. The symbiotic integration often occurs when low nutrient levels in the soil become the trigger for plants and fungi to cooperate. The fungi help plants get minerals, and in turn the fungi get sugars from the plant.

The mechanism for this interaction is that the fungi have a structure called a mycorrhiza that extends the plant's root system by becoming a functioning part of it. The extreme tip of the mycorrhiza, structures called hyphae, become extensions of the fungus capable of growing into spaces where roots cannot go and extract hard-to-get minerals such as phosphorous. In addition, the mycorrhizae protect the plant from pathogens and poisons. In generally low-nutrient landscapes, such as those found in South West Western Australia, the relationships between fungi and endemic plants become the foundation for whole ecosystems. Steve Hopper, a pioneer of this research in Australia, explains the significance of symbiosis in this type of ecosystem:

> For example, gathering and storing nutrients from highly infertile soils has
> placed a selective premium on the evolution of novel root systems . . . or
> the symbiotic partnerships with soil microorganisms, such as mycorrhizal
> fungi. The diversity of such fungi has been scarcely documented, but the

discovery of more than 300 macrofungi in the Two Peoples Bay Nature Reserve suggests a complexity equal to that of the better documented flowering plants.[17]

From even more recent discoveries within plant science comes the surprising idea that, not only do trees cooperate with fungi, they cooperate with each other to maximize chances of flourishing. For example, the health of the ecosystem is regulated by what are called "mother trees" controlling fungal networks that in turn interconnect trees of varying ages and species. The control system regulates nutrient flows to trees that need them most, such as very young ones. However, as Susan Simard has discovered, it also works to transfer information and energy from dying species to those that might continue to thrive, thus maintaining the forest as a larger system. Simard has put the case for the "wood-wide-web":

> We have learned that mother trees recognise and talk with their kin, shaping future generations. In addition, injured trees pass their legacies on to their neighbors, affecting gene regulation, defense chemistry, and resilience in the forest community. These discoveries have transformed our understanding of trees from competitive crusaders of the self to members of a connected, relating, communicating system.[18]

Similar revolutionary considerations apply in human gut function and structure. We are beginning to understand that the human body is only healthy, and ultimately kept alive, by the actions of trillions of symbiont bacteria, of many different kinds, that work with us to nurture and protect our health as well as theirs. Our "microbiome" is crucial for our physical and mental health and is regulated by the gut–brain axis via the vagus nerve. And it is not only bacteria that we live with inside our bodies. There are normally over one hundred species or types of fungi within us as well. As summarized by Bret Stetka:

> Changes in our resident microbiota and their collective genome—called the microbiome—have been linked with a wide range of diseases, from various forms of arthritis to depression. At this point scientists tend to focus on which bacterial species might hinder or maintain health.
>
> But our biota comprises a menagerie of microbes. And a growing number of researchers feel that alongside bacteria, the fungi that inhabit our bodies—or, collectively, the "mycobiome"—may also be influential in both our well-being and, at times, disease.[19]

Going along with us on the ride of life are microbiota and their microbiome, fungi and their mycobiome, and in all likelihood viruses and their virome. In addition, there are parasites of various types (e.g., our eyelash mites) that can both help and hinder optimal health. The sudden interest in microsymbiosis is largely because of the recent discoveries about connections between human health and the gut microbiome and how important they are to overall health and vitality. Microsymbiosis is even connected to the unique characteristics of place, called "terroir" by the French.[20]

We Are Holobionts

We now have a clear understanding that bacteria, trees, and humans are not individuals existing as isolated atoms in a sea of competition. The foundational idea of life as consisting of autonomous entities (organisms) in competition with each other has been shown to be fundamentally mistaken. Life consists of comingling microbiomes within larger biomes, communities within communities at ever increasing scales, otherwise known as "holobionts."[21] There is no clearly defined "inside" and "outside" of trees or humans, because, the closer we look, the more interaction between biomes we see, and the more permeable skin, leaves, and roots become. Within that permeability are not simply gases, stomata, and pores; there is a flux of bacteria, viruses, and fungi all playing a vital role in our existence as collective beings. This is more than an "entanglement" of different but independent beings; it is the sharing of a common property, called life.[22]

When we die, the loss of our life also entails the extinguishment of trillions of fellow life beings. Life is a gift held in common by healthy holobionts. The idea of life as a shared or common property of living collectives is of such scientific importance I suggest we should give it its own name. I call the shared property of life within holobionts the "biocomunen."[23] Life is precious; it is the reason for all the complexity, diversity, and relationships that animate it. Given that the whole sustainability paradigm of the past four decades failed to acknowledge this vital aspect of life, it is no wonder its principles failed to achieve "sustainable development."

The Symbioment

At both the microscopic and macroscopic scales, life has no "environment" if the environment means that which is outside of or surrounds our life.

The term "the environment" makes no sense: it is a category mistake, a product of erroneous dualistic thinking typical of Anthropocene separation. Our language and thinking have evolved on the assumption that we are not *in* the environment but are independent islands separate from it. It is time to redirect our language to reflect the reality of our total immersion in the soup of life and symbiotically animated life processes, and that means bad news for the environment.

As a consequence of the new scientific knowledge of symbiotic coexistence, I submit that we actually live within the "symbioment."[24] It is a recognition that, at its foundation, life is all about the *sumbios,* or "the living together," with each other and other types of beings. Life works with life to further life. The symbioment is now the starting point for how we can think about everything else. Along with sustainability and resilience, we must abandon the "environment" of the Anthropocene, which does nothing but perpetuate our separation from life. Nature, on Earth, as the foundation for the very possibility of life, plus organisms and holobionts, the living evolutionary experiments in life, constitute the ontology of the symbioment. Such a view has profound implications for our future, for as Lynn Margulis and Dorion Sagan have concluded:

> We have done well separating ourselves from and exploiting other organisms, but it seems unlikely such a situation can last. The reality and recurrence of symbiosis in evolution suggests we are still in an invasive, "parasitic" stage and that we must slow down, share, and reunite ourselves with other beings if we are to achieve evolutionary longevity.[25]

Sumbiocentrism

A detailed understanding of the role of symbiosis in the symbioment of life generates a new form of thinking. The root word, *sumbios,* or "living together," forms the core of a new set of concepts that have new connotations linked mainly to how humans can live in societies animated by this understanding.

As opposed to being anthropocentric, or human centered, to be "sumbiocentric" means that one is taking into account the centrality of the process of symbiosis in all of our deliberations on human affairs.[26] It requires us to give priority to the maintenance of symbiotic bonds in the total symbioment. The aim is to maximize those bonds and to hold that state of affairs in place for as long as possible. Sumbiocentrism is also an ethical position claiming that maintaining symbiotic connections, diversity, and unity within complex systems is the highest good. What is good for humans will be to live together with the richest diversity of life, to maximize the vitality and viability of interconnected life forms, including those within us.

I also think we need a new intellectual discipline, one that takes into account the way the world actually works. I call this new discipline "sumbiology."[27] Sumbiology is the systematic study of humans living together with the totality of life. Sumbiologists study life-supporting relationships between people, other biota, ecosystems, and biophysical systems in places at all levels, from the local to the global. They put flesh on the idea of interconnectedness. I am a sumbiologist. Indeed, I feel I am now a professor of sumbiology.[28] By using these new intellectual orientations, we can now enter the Symbiocene.

The Symbiocene

The first thoughts I had on the idea of a new meme called "the Symbiocene" were posted on my blog *Healthearth* in 2011. The term "Symbiocene" is from *symbiosis,* which, as detailed above, has its origins in the idea of the companionship of life (to live together). I wanted to use the profoundly important concept of symbiosis as the basis for what I think will have to be the next period (-cene) of Earth history.

The Symbiocene, as a period in the history of humanity on this Earth, will be characterized by human intelligence and praxis that replicate the symbiotic and mutually reinforcing life-reproducing forms and processes found in living systems. This period of human existence will be a positive affirmation of life, and it offers the possibility of the complete reintegration of the human body, psyche, and culture with the rest of life.[29] The path to avoiding yet more solastalgia, and other negative psychoterratic Earth emotions that damage the psyche, must take us into the Symbiocene. I will discuss the details of the Symbiocene below, but first the reasons for inventing this new epoch must be presented.

The need to create the Symbiocene was a reaction to my growing awareness that there was a concerted scientific effort to get the Anthropocene registered as a recognized period in Earth geological history. As discussed in the previous chapter, it has been claimed that anthropogenic "forcings" are now greater in global power than all the so-called natural forces at work in nature. The footprint of humanity now dominates the geology, climate, and ecology of the Earth. Not only do humans have a damaging parasitical relationship to the rest of living nature, they dominate and degrade the Earth systems vital for all life. The net result of the Anthropocene can only, in my view, end in the destruction of the Earth as a place for life in general, and human life in particular.

My negative gut reaction was a response to the realization that, irrespective of whether Earth system sciences agreed that we were in a new scientifically defined geological era, it was correct to think that humans, the *anthropos*, had come to dominate the planet. That domination, starting as "dominion" in one persistent interpretation of the Old Testament two thousand years ago, is now a hugely destructive force for the planet as a whole, even if some humans were temporarily better off because of the accumulation of power and wealth. The very idea of the Anthropocene attacked my sense of justice about the proper relationship between humans and nature, and nature and life. My gut reaction was summarized by these words:

> Many are now suggesting that we should rename this period on earth as the Anthropocene. This era could be called the Obscene, not the Anthropocene. I for one, a human, do not wish to be associated with a period in Earth's history where the dominant people in one species wipe out the foundations of life for all other humans and non-humans. I wish to be part of the "Symbiocene" where humans live in harmony with all other beings. We can do this via eco- and biomimicry and ecoindustrial economies. It's going to be hard, but it is at least thermodynamically possible. It may even be ethical and beautiful.[30]

That is the challenge we now face. The negative Earth emotions are on the ascendancy, because some humans have become hell-bent on the consumptive destruction or "creative self-destruction" of the whole world.[31] It has surprised me for some time now that while there are well-known pessimistic responses to the terraphthoric tendencies in our global culture, such as the Dark Mountain Project, there are few systematic terranascient, psychoterratic responses offering a positive vision for the future.[32] It gets worse than that, however, as the few positive efforts I can see are rapidly subverted, being turned into business-as-usual growth, easily appropriated by the forces of destruction. They prescribe technological change such as geoengineering to address problems like climate change and environmental desolation, but do not question the toxic, anti-life processes used to deliver these technologies.[33] There are of course many who see no problem at all and deny human agency in any negative change.

Living in the Symbiocene and Symbiocene Principles

The next step from the anthropocentric individualism of the Anthropocene is to apply sumbiocentric thinking and the transdisciplinary discipline of sumbiology to the task of imagining a new period in the Earth's history. The Symbiocene begins when recognition by humans of the vital interconnectedness of life becomes the material foundation for all subsequent thought, policy, and action.[34]

After a relatively short period of time (perhaps decades), there will be a point in human social development where almost every element of human culture, economy, habitat, and technology will be seamlessly reintegrated back into life cycles and processes. In order to get to that preferred state of living, I suggest that the key organizing principles of a Symbiocene society must include

- full and benign recyclability and biodegradability of all inputs and outputs;
- safe and socially just forms of clean, renewable energy;
- full and harmonious integration of human systems with biogeochemical systems at all scales;
- use of the renewable resources of place and bioregion;
- elimination of toxic waste in all aspects of production, consumption, and enterprise;
- all species, great and small, having their life interests and biocomunal properties understood and respected;
- evidence of homeostasis or heterostasis where stability is maintained and where conflict is recognized as a subset of grand-scale cooperation;
- protection of symbiotic bonds between and within species at all scales; and
- reestablishment of symbiotic bonds where they have been severed in the Anthropocene.

As all of these principles are applied, in the goodness of Symbiocene time, on the very youngest soil strata on Earth, a new, thin film of vitally organic microbiome substances will cover everything. The new "symbiofilm" will mark the proper geological commencement of the neo-Symbiocene. From that point onward, as we rapidly build the Symbiocene, that "organic" layer will become thicker and richer as it covers the multitude of sins left by the Anthropocene.

Symbiomimicry, I argue, goes beyond biomimicry (mainly concerned with copying organic form) in that the symbiotic elements of life processes are replicated in human creativity and design. In addition to mimicking the forms of life, we replicate the life processes (organic processes) that make strong and healthy the mutually beneficial association of shared life between and within different life forms.

These Symbiocene processes are not impossibly difficult to achieve, even given the current state of human science, folk knowledge, and technology. If we put real effort into Symbiocene science, citizen science, indigenous knowledge, and technology, use the very best of "biomimicry" and the new domain of "symbiomimicry," we could get well into Symbiocene living in only decades.[35]

The Symbiocene entails a real revolution, because, for example, we must replace every single toxic and polluting Anthropocene artifact with an equivalent (if possible) benign Symbiocene "sumbiofact." With over 7 billion people on the planet now, there can no longer be tolerance of anthropogenically produced toxic-to-life substances used in human economic production and enterprise. There cannot be anything less than the complete adoption of the Symbiocene principles.

Those who want innovation to drive a new economy that is productive within the associations of life now have a new narrative or meme to stimulate their creativity. The core idea of the Symbiocene is pro-growth. However, it is not pro-parasitic or terraphthoric growth, which is tierracide (Earth murder). It is the negation of these, and their replacement with a form of growth that is consistent with symbiosis, a foundation of life on Earth as we know it. Terranascient growth is good, and we need as much of it as we can create.

There can be, however, no "green revolution" (ecomodernism) based on existing technologies or trickle-down "natural capital" generation. Rather, it will be a revolution based on the rapid and complete transition from a polluting and consuming society to one that produces everything via nonpolluting, symbiotically active means. Instead of retreating into primitivism or atavism, the Symbiocene requires a massive surge in innovation and creativity. The surge will not be generated by the Left or Right in political terms, nor will it be socialist or capitalist.

Gone will be "sustainability" and terms like "sustainable development." In their place I suggest difficult-to-corrupt terms such as "sumbiosity" and "sumbiosic development."[36] There will be active invention and creation on the part of humans to achieve and conserve a state of sumbiosity. It will be within sumbiosity where we will see humans and other life forms living together in mutually supporting relationships. Such a state of affairs will be new and will be based on positive Earth emotions that see shared life as the highest value.

The policy realm will also have to be revolutionized by sumbiocentric thinking. The model for social systems is not nature in general, it is life, and how life is possible (and desirable) within the matrix of an ethically indifferent nature.[37] From this point onward, there is nothing to stop the new conceptual framework from evolving greater complexity and delivering new terms for those that have become redundant and dangerous. We can even create the conceptual framework for a new overarching philosophy of life.[38]

Sumbiocracy

The emergence of new forms of governance that reflect the new symbiotic understanding of the shared project of life is now a possibility. Sumbiocracy is rule for the Earth, by the Earth, so that we might all live together.[39] Sumbiocracy is, after Abraham Lincoln, "government of the Earth, by the people of the Earth, for the Earth, so that the Earth shall not perish."

"Sumbiocracy" I define as a form of cooperative rule, determined by the type and totality of mutually beneficial or benign relationships, in a given sociobiological system.[40] Sumbiocracy is a form of government where humans govern for all the reciprocal relationships of the Earth at all scales, from local to global. Organic form (all biodiversity including humans) and organic processes (symbiotically connected ecosystems and Earth systems) are primary in this new form of government.

If the processes that nurture and maintain symbiosis within ecosystems, biomes, and organisms are identified, protected, and conserved, species within such healthy systems will also flourish. We need to further improve the biased, anthropocentric notion of democracy (from *demos*, people) with, say, a deep ecology "council of all beings" approach where the interests of species are represented in decision-making structures by well-meaning humans.[41] While species are important, as we have seen, it is the shared life between species that must also come to the fore. Therefore we need to elect people to govern who understand and affirm life-supporting organic form, *process,* and relationships, such that they can deliberate on creative proposals from humans. Sumbiocracy will be governance for symbiotic relationships between and within species, such as between plants and fungi, humans and bacteria, humans and other animals. Deliberative sumbiocracy will be the appropriate form of governance and decision-making for the Symbiocene at all scales, from local to global.

Sumbiocracy requires those who govern (sumbiocrats) to have a thorough understanding of the symbiotic, life-sharing interrelationships that enable them to function. In order to "live together," humans must exercise their intelligence and power to assist in creating justice or overall harmony in a community of life interests. Within a sumbiocracy, sumbiocrats must ponder what kind of mutualistic development is permissible to enable living together to be fulfilled. Breaking symbiotic connections at all scales must be avoided, and it will be good governance to identify vital symbiotic life connections and to ensure protection for them to remain in place.

Governance by scientifically and traditionally informed humans (including citizen scientists) at *all* places and *all* scales identifies the interconnections between elements of complex systems before they commit to action that impacts ecosystem health. We must also remember that place is critical for effective sumbiocracy, since those with close and intimate ties to particular places or bioregions are in the best position to know their place and to make decisions about its health and vitality.

Rights?

While it might seem initially counterproductive, the extension of "rights" to nonhumans in an effort to bring elements of ecosystems into the circle of human ethico-legal protection will not be in the spirit of the sumbios. This is because "rights" within Western thinking have their origin in exploitative and manipulative modes of human decision-making, arising from the need for secular, private wealth and property to be protected from the privilege of institutions such as the church and the Crown.[42] The possession of rights depends on the idea of the autonomous individual as the bearer of rights in a contestable social context. Moreover, a hierarchy of rights emerges where, for example, males and their power structures give preference to "masculinist rights" over all others.[43] Rights are wrongs, and we do not need them in a nonhierarchical world of permeable and porous intersections of life interests. The only place where the language of rights fits within the sumbios framework is the idea of a "right of passage." However, this right is confined mainly to ships and shipping in the navigation through international and national waters.

Within what can be called "sumbioethics," Symbiocene principles can be applied to the pursuit of a good life. In addition, there will be a need for a new concept for "rights," one that takes into account symbiotic interconnectedness within the symbioment, the human sumbios, and the human biome. I suggest a replacement of the rights concept be called "ghehds," where instead of a hierarchy of competing rights, assuming autonomous individuals or entities in a contested domain, ghehds are the entitlements of vagility, passage, movement, and flow within organically and symbiotically unified wholes. The good of the whole is guaranteed by the protection of ghehds that connect and hold things together. "Rights" assume division and exclusion; "ghehds" assume unity and inclusion.[44]

The idea of ghehds also invites new approaches to issues of injustice where, for example, land was appropriated from indigenous peoples and then must be fought for to be returned to them in Western court systems based on competing claims about land "rights." Decolonization and a return of indigenous people to a living, symbiotic relation to their own land is also a form of ghehds-based radical justice. I see ghehds as similar in function to the song lines of Australian Aboriginal people. They tell people how to live successfully on land.

The Impact of the Symbiocene

While in its very early stages of development in my thinking, the Symbiocene as an idea has already found a few homes. In the development of a jurisprudence system for Earth justice, the United Nations has endorsed my idea, suggesting that

> current approaches to the Anthropocene epoch that focus on human impacts on the Earth's biogeochemistry need to be expanded. Concepts such as the Symbiocene, an era when human action, culture and enterprise would nurture the mutual interdependence of the greater community and promote the health of all ecosystems, are more promising and solution-oriented.[45]

In the context of global health, the Symbiocene concept has also been taken up by health academics and professionals. In order to maximize human, ecosystem, and planetary health, the need for humans to repair and maintain symbiotic associations at all scales of life will be a major primary health care responsibility. Symbiocene health will ensure good human health, as it will for all other organisms on this most rare of planets that sustains life. Contemporary public health experts are now supportive of this biophilic and sumbiophilic direction:

> In sum, the symbiocene represents an era where education and awareness of the benefits of empathic and emotionally intelligent citizens, institutions, and systems will be prioritized for societal good. The driver of this process will be biophilic science. Such a science describes investigations with an aim toward the promotion of life, human, and total species welfare in mind. Biophilic science understands that allostatic load, dysbiosis, mental distress, microbiome diversity, living wages, and environmental toxin exposure do not occur equally throughout society; it also understands that positive psychological assets are not on a level SES playing field. . . . Biophilic science in the symbiocene seeks to develop ways to curb health-corrosive exposures and identify assets of resilience, while at the same time prioritizing its activities toward those with the greatest need.[46]

As with solastalgia, contemporary artists have been inspired by the Symbiocene to negate the Anthropocene. The artists Jenny Brown in Australia and Cathy Fitzgerald in Ireland have both delivered powerful responses to this relatively new idea.[47] Art theorists are also seeing the connections between the defeat of solastalgia and the rise of the Symbiocene.[48]

The idea of the Symbiocene is now being applied in and has the potential to be a stimulus for many new directions in the environmental humanities (sumbiomental humanities) and the applied sciences and professions. Rod Giblett, my former academic colleague in Western Australia, has made a great start by arguing for a new bioregional approach to embracing the living Earth in the Symbiocene. I hope he is the first of many more who will use the new meme to create a positive and optimistic vision of the future.[49]

Some Antecedents of the Symbiocene

How utopian or blindly optimistic is the idea of the Symbiocene? Is it an atavistic or futuristic fantasy? At one level we could argue that for the bulk of time that *Homo sapiens* has been a species on the face of the Earth, we were within a proto-Symbiocene state, as nearly all enterprise satisfied the Symbiocene principles I outlined above. It was only at the point of the Industrial

Revolution that our own development as a global species began to deviate from the matrix of the rest of life. Pollution of the atmosphere, and the use of nonbiodegradable and toxic chemicals in industrial processes began slowly but have escalated since the time of my birth in 1953. Even agriculture was, by default, "organic" prior to the Second World War, and was changed rapidly by the use of fossil fuels for energy, fossil fuel–based fertilizer, plus mass application of agricultural chemicals. The "great acceleration" of industry, agriculture, and technology in the second half of the twentieth century was achieved without much attention, if any, to the centrality of symbiosis as a foundation for life.

However, there have been many writers and thinkers who have championed the idea that humans do have the capacity to live in symbiotic harmony or stability with each other and the rest of life. I consider Peter Kropotkin, in his *Mutual Aid as a Factor in Evolution* (1901), an important thinker who saw much more than greed and selfishness in animal and human nature.[50] Many modern environmental thinkers and writers have reached the same conclusion, with the need to express this cooperative side of human nature in ethical and policy principles. Writers as diverse as Lynn Margulis, Dorion Sagan, Murray Bookchin, James Lovelock, Donna Haraway, and Tim Morton have been or are champions of the symbiotic revolution that has transformed the way we think about ourselves and our relationships to other beings.[51] Donna Haraway, in particular, has given generously of the creative imagination needed to take symbiosis out of bioscience obscurity and into the environmental humanities in the twenty-first century. Echoing my own revulsion about the Anthropocene, she makes a case for its replacement as the ambivalent "Chuthulucene," where the great drama of life and death is played out but without the terraphthoric forces of the Anthropocene.[52] I am in complete agreement with her when she asks, "Why is it that the epochal name of the Anthropos imposed itself at just the time when understandings and knowledge practices about and within symbiogenesis and sympoietics are wildly and wonderfully available and generative in all the humusities, including non-colonizing arts, sciences, and politics?"[53]

In order to build the case for sumbiosic thinking, I wish to highlight once again the contribution of Elyne Mitchell, the Australian writer-scholar who has influenced my work. Further, in working through some of the issues Mitchell writes about, the sumbiosity of Australian Aboriginal people will be examined.

Soil and Civilization

Elyne Mitchell, while writing in Australia during the Second World War, drew on early ecological thinking to expound a philosophy of life. Mitchell published her book, *Soil and Civilization*, in 1946, before the availability of Aldo Leopold's famous *A Sand County Almanac* in 1949. In her modest book, Mitchell builds on the early scientific knowledge of symbiosis as a foundation for life and weaves a narrative around the need for Australians to apply this knowledge, basic science, and the "Dreaming" of the traditional Aboriginal people of Australia to build a new foundation for a viable future. Needless to say, prior to and during the Second World War there were very few people writing about symbiosis as important for a good human-nature relationship, or of Indigenous Australians as having something positive to contribute to a sumbiosic future for Australia.

While there is little material on Indigenous people and culture in Mitchell's books, there is enough to build a more inclusive and integrative philosophy of life that engages an Indigenous worldview. I am also acutely aware that Mitchell repeated beliefs about Aboriginal Australians that could be considered racist. She was a white, Christian, rural, self-educated woman, it was the 1940s, and her reading audience was "white Australia." Yet perhaps in defiance of the public orthodoxy of the time, she wanted to acknowledge as a basic fact that, prior to colonization by Europeans, "the nomadic aboriginal was part of the natural balance of the continent."[54] She further observed that the colonizers had destroyed that balance in less than two hundred years.[55]

However, there is, in her references to symbiotic science and Indigenous culture, sufficient material to tell us something important about the relationship between nature and humans, culture and nature. Mitchell saw a tragic mismatch between the European mind and the biophysical reality of Australia in 1788, and *Soil and Civilization* was her response to the spiritual and physical degradation of the Australian continent. She suggested:

> The natural laws of the undiscovered Australia were incomprehensible to minds moulded in Western Europe. Yet that this land may survive as living earth we must learn to understand the balance that existed when Australia contained only the nomadic aborigines and the slow-breeding marsupials, and condition our relationship with the land by what we learn.[56]

Mitchell mentions the "nomadic aborigines," seeming to accept the universality of the nomadic nature of an Aboriginal hunting and gathering lifestyle and its cultural base. She writes that "aborigines, who collected grass to grind for baking, sowed no seeds, and no vegetation myths entered into their experience."[57] No doubt her views would have been seriously challenged by the evidence and ideas presented by contemporary Australian writers such as Bill Gammage in *The Biggest Estate on Earth* and Bruce Pascoe in *Dark Emu*.[58] However, the idea that Aboriginal people cultivated the soil, used tools such as hoes, and harvested grain that was actively planted by them would have enhanced the thesis of her book, that an agriculture sensitive to the unique Australian conditions was both desirable and possible.

In addition, Mitchell noted that, while Aboriginal culture did not have specific Greek-type "chthonian Gods" (connected to the underworld, soil, and agriculture), it was located within what she describes as a "boundless space" known as "the matrix of the Alcheringa, the dreamland of the aborigines." Here, she observed the possible unity of "this soil" and "this cosmic vastness" leading to a "permanent culture . . . of small or larger numbers, as the possibilities of the land allow."[59] Such a unity also required a "fusion of ancient wisdoms with all that modern science can discover."[60] The ideas of permaculture, ancient wisdom, the value of science, and limits to growth were presented to the world in 1946 by a wise woman pastoralist in Australia.

The unusual use of the term "dreamland" deserves some comment, as "land" and "place" figure prominently in all of Mitchell's formal writing. I can only assume that she had available to her early anthropological accounts (before World War Two) of the Dreaming and what the "dreamland" might signify. I do not know what her actual sources were. However, A. P. Elkin published his popular book *The Australian Aborigines* in 1938, and references to the Dreaming lie within it. The debate about the origins of this term have been well discussed in the recent literature, so I do not need to cover this ground again.[61] Elkin gives a summary of what, for him, were the core philosophical dimensions of the Dreaming:

> The Aborigines have a view of life and nature, more or less logical and systematic, granted its animistic premises, which is a lantern to their feet and a guide to their paths, as they pass from birth to death and beyond. It is spiritual, totemistic and historical in nature, expressing the central facts of human personality, of man's intimate relationship to nature, and his tie to the past, all of which is carried over in a belief that personality is beyond space and time.[62]

Such a total worldview created what I understand as a type of proto-Symbiocene. The merging of culture and nature meant that there was very little evidence of a uniquely human impact on the land, as a culture symbiotically connected to the land leaves very few material traces.[63] Stanner also reached such a conclusion. He argues:

> They are, of course, nomads—hunters and foragers who grow nothing, build little, and stay nowhere long. They make almost no physical mark on the environment. Even in areas which are still inhabited, it takes a knowledgeable eye to detect their recent presence. Within a matter of weeks, the roughly cleared camp-sites may be erased by sun, rain and wind. After a year or two there may be nothing to suggest that the country was ever inhabited. Until one stumbles on a few old flint-tools, a stone quarry, a shell midden, a rock painting, or something of the kind, one may think that the land had never known the touch of man.[64]

Again, Stanner may well have changed his mind on the issue of agriculture and evidence of semi-permanent settlements in parts of Australia, yet his account of the Dreaming does connect to my notion of the Symbiocene. The integration of culture and nature produces a symbiotic and benign result. There is nothing produced by humans in the air, water, soil, and landscape that cannot safely be assimilated back into the vital matrix of life. There is symbolic and material kinship between humans and other life forms, and they are mutually supportive. What nowadays would be called "environmental ethics" were built over thousands of years by Aboriginal people, to maintain and protect the special relationships between people and place (country) all over Australia. The Symbiocene principles were all in evidence in traditional Aboriginal society.[65]

The politics of traditional Aboriginal society also ran counter to Hobbesian notions of scarcity, and the ideas of perpetual conflict and aggressive colonialism.[66] Conflict was minimized because people were so immersed in particular areas of land and the resources contained within it. It made no psychic or cultural sense to take over the land of others. Individual identity (their personal Dreaming) was tied to intimate knowledge of their land. A massive proportion of language acquisition, education, and socialization of children was symbiomental education. Trade and social intercourse were governed in such a way as to minimize conflict and maximize cooperation.[67] The deeper the symbiosis between people and the land, the deeper the lasting peace between different geographically distinct peoples. I will return to this theme in the final chapter.[68]

It is my view that Mitchell could see the urgent need for a version of this kind of cosmology to enter into Australian colonial life and all of its enterprises. In agriculture, where she could see the soil blowing away and plant life receding in vitality, she was especially critical of a culture divorced from the biophysical reality of symbiosis and one lacking related spiritual and ethical constraints on greed and selfishness.

Selectively researching alternatives in the agricultural context, Mitchell focused on agricultural experiments that built soil fertility, based on an assumption and knowledge of the interdependencies between "soil, plant and animal." She argued that

> The principle underlying all these experiments . . . was this *living, organic symbiosis* which goes deeper and leads far beyond their actual quest for physical health in human beings, stock and plants. In the proof of biological interdependence, there is evidence of a universal pattern which is an outward form of the rhythm to which life moves.[69]

Mitchell encouraged Australians to have a "vital love of the sense of being" by living in unity and harmony with the wider cosmos and also the phenology of place. An early champion of bioregionalism, she encouraged a cultural reunion with the "essential" Australia via the acquisition of a new "land sense."[70] She also encouraged a "real love of the land—a love of the universal Australian earth and an intense love (endemophilia) of the particular place from which each individual comes."[71] Her vision of such a vital reunion even involved a symbiotically inspired cultural revolution: "The courageous decision to build civilization into a *symbiosis* with a revitalised world, possessing stable, healthy soil, clear streams, unburnt forests and dust-free air, with people working on the land, close to reality, would bring new balance and new life."[72]

These ideas of Mitchell's are early attempts within the Australian context to encourage humans to move away from a form of civilization that destroys its own biophysical foundations and, in the process, destroys psychic, physical, and cultural integrity. They present a vision of a symbiotic civilization that must be one of the first in any context to be presented in a public forum. Long before later champions of symbiosis such as Lynn Margulis and Gregory Bateson made similar connections between biophysical health and mental health, Mitchell, in 1946, made a strong case for just such a connection.[73] As mentioned in chapter 2, on the link between the health of the land and psychological health in people, she wrote:

> But no time or nation will produce genius if there is a steady decline away from the integral unity of man and the earth. The break in this unity is swiftly apparent in the lack of "wholeness" in the individual person. Divorced from his roots, man loses his psychic stability.[74]

I connected Elyne Mitchell's notion of "psychic stability" to the concept of solastalgia from my earliest thinking about land-psyche relationships. Global psychic instability was manifested in a world at war, yet Mitchell was also prescient in seeing that a significant factor in geopolitical instability and war was that "predatory civilisations" arose based on the need for constant colonization to exploit new resources. Once this process goes global, she mused that "it will not be in just one small area that the spirit of man will be extinguished, but almost throughout the whole world."[75] I return to the loss of "spirit" in the next chapter.

Similar connections have been made by more contemporary thinkers on the condition of humanity. The late Lynn Margulis, mentioned above, also related the state of the psyche with the state of the biophysical world. She and Bruce Scofield argued that "the psychological discontent of civilized people emerges directly from isolation, a chronic physical *dissociation* of humans from the rest of the biosphere, including fellow humans."[76]

The Earth was busy evolving as a "symbiotic planet," while humans have attempted, over the last 300 years, to socially evolve in isolation and in defiance of the direction taken by the rest of life. The negative implications of physical *dissociation* from the rest of nature and life are now obvious. Such an observation is critical to our future on Earth. If we keep heading in the anti-symbiotic direction (dysbiosis), then the psyche apocalypse and the eco-apocalypse will be simultaneously upon us, and the dystopian turn in film and fiction will become a lived reality.

Symbiocene Summary

Australian Aboriginal people developed a social-symbiotic relationship with the elements of their "country" over a period now accepted to be between

sixty-five to eighty thousand years of continuous occupancy of the Australian continent and Tasmania.[77] The emphasis was on deep integration and
intertwining with the elements of the symbioment, and this is the basis of
a peaceful coexistence with other beings and, most important, the basis for
peaceful coexistence with other groups of human beings who lived in and
related to their own "country." These diverse relationships with other beings, built over deep time, are not mere "entanglements." They reflect the
biocomunen, the intricate life-sharing living arrangements between different
types of beings at various scales of existence.

There is also no room here for a crude form of xenophobia, but a considered respect for other tribes to look after their own patch and live by the
rules of their own Dreaming. Of course, there is interaction and maybe at
times conflict, but there is an overall pattern of symbiosis with profound
implications for how different groups of people can live together in large
spaces such as continental Australia. There is a way of being human that
enables peaceful coexistence between the "song lines" of interdependent
"countries." A pattern of ancient symbiosis was created in Aboriginal social
society. It can be replicated by empathetic humans in the Symbiocene in the
form of a confederation of sumbiocracies.

Positive Psychoterratic Conditions

In the Anthropocene, negative psychoterratic emotional states begin to pre-
dominate as the biophysical and built environments are desolated. In the
Symbiocene, positive psychoterratic emotional states are nurtured as the
biophysical and built symbioment are reunited. Those who have not yet
been seduced by the Anthropocene still have within them remnants of previously unnamed positive Earth emotions. I will put the case below that
humans are innately or instinctively empathetic to life, and desire it to be
preserved and to continue. When compared to nonexistence (death), life
seems incredibly good.

There are many life-affirming belief systems throughout the world, with
some religious traditions, such as Jainism, extremist in their protection of
all forms of life. Albert Schweitzer, with his *veneratio vitae*, or "reverence
for life," built a whole system of ethics based solely on this affirmation. He
argued in *Civilization and Ethics*:

Ethics grow out of the same root as world- and life-affirmation, for ethics, too, are nothing but reverence for life. That is what gives me the fundamental principle of morality, namely, that good consists in maintaining, promoting, and enhancing life, and that destroying, injuring, and limiting life are evil.[78]

In table 2, I list the positive psychoterratic emotions developed in the literature over the last few decades. By no means complete, the list has the potential to expand as additions are made by those who identify and champion a unique positive Earth emotion or feeling. It will bring me great joy to see the list expand in the future.

Biophilia and Other Earth Philias

In the 1960s, Erich Fromm further developed the Schweitzerian idea of life as intrinsically valuable by drawing the contrast between the love of life and the love of death. As indicated in the previous chapter, "necrophilia," or the love of death and destruction, can be contrasted with "biophilia," or the love of life. In the *Heart of Man*, Fromm develops the idea of biophilia in the context of personality development:

> The full unfolding of biophilia is to be found in the productive orientation. The person who fully loves life is attracted by the process of life and growth in all spheres. He prefers to construct rather than to retain. . . . Biophilic ethics have their own principle of good and evil. Good is all that serves life, evil is that which serves death. Good is reverence for life, all that enhances life, growth, unfolding. Evil is all that stifles life, narrows it down, cuts it into pieces.[79]

Table 2. Origin of positive psychoterratic states

Positive state	Origin and year first used
Topophilia	Auden, 1947; Tuan, 1974
Biophilia	Fromm, 1964; Wilson, 1984
Ecophilia	Sobel, 1996
Soliphilia	Albrecht, 2009
Eutierria	Albrecht, 2010
Endemophilia	Albrecht, 2010
Tierraphilia	Albrecht, 2014
Sumbiophilia	Albrecht, 2016

Fromm's pioneering concept of biophilia conjoins the love of humanity with love of life and nature in a way that anticipates many themes in late twentieth-century and early twenty-first century environmental writing. In his "Humanist Credo," published in *On Being Human,* he linked his understanding of biophilia to a comprehensive ethic: "I believe that the man choosing progress can find a new unity through the development of all his human forces, which are produced in three orientations. These can be presented separately or together: biophilia, love for humanity and nature, and independence and freedom."[80]

Fromm's biophilia is certainly a positive psychoterratic state. He argues, however, for no deep, mutually beneficial, symbiotic view of the cooperation present within life. To develop his life-nature-based concept of biophilia even further, into a more empathetic and ecological understanding of life, we might wish also to talk about "ecophilia," or the love of the total ecosystem within which one is located. In 1996 the environmental educator David Sobel pioneered this view in his opposition to the negative psychoterratic condition of ecophobia. He wrote: "I propose that there are healthy ways to foster environmentally aware, empowered students. We can cure the malaise of ecophobia with ecophilia—supporting children's biological tendency to bond with the natural world."[81]

"Biophilia," as a psychoterratic term for a positive Earth emotion, was also used by E. O. Wilson decades after its creation by Fromm. Wilson defines biophilia as "the innate tendency to focus on life and lifelike processes."[82] However, he uses the term in a way that is genetically or biologically rooted, as distinct from Fromm's neo-Freudian personality-development perspective. Wilson has argued for a "deep conservation ethic" based on an innate biological affiliation with all other organisms, as a counter to destructive impulses such as necrophilia. It has always troubled me that biophilia must be a very weak innate emotion, because modern humans seem so easily to slip into values and actions destructive of ecosystems and their life forms. Besides, it is hard to focus on life and lifelike processes, when they are largely invisible to us. The "microcosmos," as Margulis and Sagan call it, is an invisible world where biophilia has no object for its attention. This is philosophically important.[83]

Wilson, in *Half Earth,* does acknowledge the huge expansion of our understanding of symbiosis in the living world. He refers to the "bacterial gardening" within us that is needed to keep our bodies healthy, and goes on to suggest:

The gardens growing within human beings and other animals are typical of complex ecosystems everywhere, inside and out. The overall number of kinds of microbiomes dwelling in animals and plants worldwide remains entirely unknown, but it must be enormous. . . . It is obvious that symbiotic microbiology, which encompasses these systems, has emerged as an exciting frontier of science, and will remain so for decades to come.[84]

Despite his awareness of symbiosis, Wilson does not take his conservation ethic beyond saving half the world's surface area for nature. Such an effort would be useless if, for example, climate warming causes breaks in the vital symbiotic linkages between Earth-building symbionts. He diagnoses the world as in "desperate condition," yet still clings to biophilia as "the key to doing no further harm to the biosphere."[85] Wilson can see the advance of science in the domain of symbiosis but does not go far enough in extending and strengthening biophilia as a moral or educative precept.[86] As if to acknowledge this depressing outcome, Wilson names the next possible era in human history the "Eremocene," or the "Age of Loneliness," where there is nothing on Earth except people, agriculture, and domesticated animals.[87] We need to go beyond biophilia and even ecophilia in order to capture the essence of the symbiotic revolution in the biosciences.

Sumbiophilia

I now put the case for "sumbiophilia," or the love of living together, as an addition to biophilia and ecophilia, to capture this new understanding that, not only is life connected in ecosystems, it is also interconnected within organisms.[88] Clearly, previous generations of humans could not have had any knowledge that their bodies are home to trillions of bacteria and fungi, so the microscopic symbiont world is new to us. Despite this, coevolution between the microbiome and human bodies must sit within us as some kind of "unconscious," instinctual knowledge. Our food preferences, for example, just might be both a vehicle for nutrition plus a selection for probiotic gut help. We eat to feed our good bacteria. You are what your bacteria eat, even to the level of moods and emotions. I mentioned the importance of terroir above, and it too is a logical and material home for the emotion of sumbiophilia.

Once the story of our symbiont interconnections is more fully understood, humans will have a new knowledge base on which to build their values and their actions. To maximize health and keep symbiogenesis going (in addition to simple evolution by natural selection), sumbiophilia will be an outcome of new knowledge. Education, plus whatever relictual element of innate or unconscious biophilia, ecophilia, and sumbiophilia we have within us, will propel change toward the Symbiocene.

If I am correct, then exiting the Anthropocene and entering the Symbiocene will be a deeply satisfying experience for most humans (and their gut bacteria). Our instinctual love of life (macro, meso, and micro) and lifelike forms can prevail over eco-necrophilia and possible ecocide. At the global scale, "tierraphilia," or the love of the Earth, is a logical extension of sumbiophilia.[89] For all we know, this Earth might be the only place in the cosmos that is a home for life. It could be so special that something bigger than love might be needed to give homage to our home.

Topophilia

The concept of "topophilia" was first used by the poet W. H. Auden, in 1947, to describe the attention given to the love of particular and peculiar places, as evident in the poetry of John Betjeman.[90] The neologism combines *topos* (place) with *philia* (love), hence "love of place." In Auden's mind, topophilia "has little in common with nature love. Wild or unhumanised nature holds no charms for the average topophil because it is lacking in history; (the exception which proves the rule is the geological topophil)."[91] As I argued in chapter 2, a person can have solastalgia for aspects of the built environment, so Auden's idea of topophilia is a useful positive psychoterratic emotion.

The geographer Yi-Fu Tuan added to the scope of the concept of "topophilia" by including the nonbuilt or natural environment, as well as the built environment, in the realm of "love of landscape." In defining topophilia, he highlighted the suite of human "affective ties with the material environment."[92] Tuan argued that, on most occasions, topophilia is felt as a mild human experience, an aesthetic expression of joy about connection to landscape and place. But he said that it can become more powerful when human emotions or cultural values are "carried" by the environment. In his 1974 book, *Topophilia*, he acknowledged the work of the anthropologist Ted Strehlow, who provided insight into the depth of positive place attachment held by Aboriginal Australians and what happens when place attachment is severed.[93] He acknowledges that topophilia can be a powerful human emotion for humans who are closely connected to the land.

If we accept that love of landscape and place can be a powerful emotion, especially for indigenous people and people who live closely to the land and soil, then a lived experience of the chronic desolation of that landscape or place would be an equally powerful emotion and psychic state. Under climate warming we have seen, for example, the regular occurrence of human-enhanced natural disasters from massive storms in places like New Orleans, Houston, New York, Mumbai, and many other urban locations worldwide. People who love their landscapes—whether they are cities, rural villages, rivers, or forests—are now having that love torn away from them. Topophilia is replaced by solastalgia and tierratrauma.

Soliphilia

As I argued in chapter 3 with respect to solastalgia being an emotion that could be reversed and alleviated, I also realized that this negative emotion could be actively fought against within social and political contexts. As solastalgia can be imposed by powerful government and corporate forces, it can be resisted by confronting those very same forces.

Rather than giving in to the melancholia of solastalgia by resorting to forms of escapism or defeatism, a positive response in the form of community involvement in addressing both the cause of the issue and rehabilitation of the resultant damage is both possible and desirable. By ceasing to see the impact of environmental damage as strictly personal rather than an issue held in common with other people, there is empowerment. No longer is the issue one that "blames the victim" or one where the victims blame themselves. The negative emotion serves now as the foundation for a positive one.

I set about creating a political neologism to explain the process of converting the negative into a positive. As most "environmental" and "development" issues are political and sit within the orthodox left-right political spectrum, I thought it would be useful to have a neutral political term for the process of fighting solastalgia. It is a pity that the concept of "solidarity" has been strongly identified with the Left in politics, as it could be usefully applied to the politics of removing the causes of solastalgia.[94] It is bad enough that we still engage in the orthodox left-right fight about who should own ecocidal industrial society; we should not fight politically over how to defeat solastalgia. In 2009 I created the concept of "soliphilia" to provide a cultural and political concept that will help negate global dread and solastalgia.[95]

Soliphilia is the love of the totality of our place relationships, and a willingness to accept the political responsibility for protecting and conserving them at all scales. The concept has its origins in the French *solidaire* (interdependent) and the Latin *solidus* (solid or whole), and the love of one's fellow citizens and neighbors implied by the Greek *philia*. Soliphilia is manifest in the interdependent solidarity, and the wholeness or unity needed between people, to overcome the alienation and disempowerment present in contemporary political decision-making about the symbioment. Soliphilia introduces the notion of political commitment to the saving of loved places at all scales, from the micro, local, and to the global. This "philia" is a culturally and politically inspired addition to the other philias that have been created to give positive biological and geographical conceptions of connectivity and place. While only in existence since 2009, this concept has already been discussed in feature articles on ecopsychology and politics that have global reach.[96] It has even inspired a few artworks.[97]

Soliphilia is now added to love of life and landscape to give us the love of the whole, and the solidarity needed between humans to keep healthy and strong that which we all hold in common. The relevance of soliphilia lies in the hard work that people must do, all over the planet, to save those places that are under terraphthoric assault. In order to negate all the "algias," or forces that cause sickness and extinction, we require a positive love of place, expressed as a fully committed politics and as a powerful ethos, or way of life. Soliphilia goes beyond the left-right politics of the control and ownership of cancerous industrial growth, and provides a universal motivation to achieve sumbiosity through new ways of symbiotic living that are life-affirming. Soliphilia and sumbiocracy together provide a new foundation for governance that respects symbiosis.

SOLIPHILIA IN WESTERN AUSTRALIA

In Australia, I have studied the Margaret River region of South West Western Australia for its soliphilic properties. The area sits between two capes on the Indian Ocean and is often called the Cape to Cape region. Margaret River is a case study par excellence for soliphilia. Here, in the year 2009, a proposal for an underground coal mine galvanized people in the region to defend their terroir. The region is famous for its mild climate, scenery, forests, beaches, vineyards, and cultural activities, so the very idea of a coal mine sat like an ugly sore within its sense of place. The people interviewed were outraged by the proposal because, as one interviewee put it, "If there was anywhere better on earth, I would already be there."[98] Others witnessed their love of place (topophilia) through expressions such as "This is God's own country" and confirmed the pleasure they derive from networking with others to maintain what the entrepreneurs call "brand Margaret River." Just one negative change to the psychoterratic nexus, and the whole edifice is threatened and could fail.

The Cape to Cape region, in the colonial phase of its environmental history, has attracted humans with identifiable values and attitudes that are alternative to the mainstream. Surfers, hippies, artists, and retired doctors live within the same geographical space. Further, these values and attitudes are centered on a strong (endemic) sense of place. Over time, the colonists of this region have built a network of mutually supportive enterprises based on the resources and characteristics of the biophysical region. And they have done really well. The wines from this region are now world famous, and many family vineyards, such as Cullen Wines, have pioneered outstanding organic and biodynamic viticulture. The internationally renowned wine writer Jancis Robinson said that "Margaret River is the closest thing to paradise of any wine region I have visited in my extensive search for knowledge."[99]

Along with the wine comes local and regional produce that are used in restaurants to support the cultural terroir. The participants in this bioregional cooperative human endeavor could be considered "cultural symbionts" as they receive benefits from others in the region and, in turn, give benefits back to them. The Indigenous people of the area are also working on their reconnection, after colonial disintegration, of culture and place within the Wardan Aboriginal Cultural Centre.

In addition to the vineyards, there are art galleries set in wildflower bushland, as the region is home to many artists. The surfing beaches are world famous for the big waves that roll in from the Indian Ocean. I was lucky enough, while visiting the region, to find a local wood craftsman who specialized in the native timbers and turned them into fine furniture. My coffee table is a constant reminder of the fine quality of the Jarrah tree with its red inner glow and endemic, ebony-like hardness.

Although with a diverse group of citizens, the research I conducted confirmed that there is unity in diversity, especially when a common enemy such as coal mining or gas fracking is concerned. While not technically within the Symbiocene, place-attachment values that are in harmony with the biophysical base of the region make for a good start. When threatened, these people react with a rapid bioregional response and form alliances that cross old political divisions to protect a unique place.

In 2012, the prodevelopment conservative state government banned all further exploration and development of coal reserves within the region. The Labor government, in control of the state in 2017, has also banned gas fracking in the region. However, the pressure to extract gas and coal continues to threaten the area. Political pressure from the federal government to exploit these resources, in return for a greater share of federal revenue, is held out as both a carrot and stick approach to development. Don't develop these "resources," and the state will get less revenue from the federal government. It demonstrates that more concentrated political power at local and regional levels is needed to protect what remains of ecological and cultural bioregionalism.

I am certain that, at the local and regional level, soliphilia will continue to be a political force driving people to fight all forms of development unsympathetic to the cultural mutualism that maintains the health and vitality of the region right now. Sadly, as we saw in chapter 2, in the case of the people of the Upper Hunter, especially in Bulga, soliphilia was not enough to save the day. However, I remain stubbornly optimistic that soliphilia will, in the long run, win the fight.

In the areas around NSW where gas fracking has been an issue, a soliphilia-based protest group called the Knitting Nanas Against Gas (KNAG) has been formed. It is a form of pacifist protest where mature women sit together and knit symbols of their opposition to gas fracking and, at times, open-cast coal mining. We now see such mature-age protest groups all over the world; however, it is mainly women who are the driving forces behind such activism. The region of Gloucester is also under threat from two open-cast coal mines and a proposed new mine on its city boundary. The citizens (including their children who oppose the coal mines and the gas fracking) provide another example of soliphilia-based social and political protest and collaboration. I live in hope that "Grandfathers against Terraphthora" will be formed very soon. "Blockadia" will just keep getting larger and stronger as it is fueled by more soliphilia worldwide.[100]

Endemophilia

I mentioned this term in chapter 1, my sumbiography, as it is central to my love of place. I created "endemophilia" to define and cover the particular love of the locally and regionally distinctive in the people of a place.[101] Endemophilia is based on the English word "endemic," which in turn is based on the French word *endémique* and has the Greek roots *endēmia* (a dwelling in) and *endēmos* (native in the *demos,* or people) and *philia* (love).

Once a person realizes that the landscape they have before them is not replicated in even a general way elsewhere in their country or on their continent or even in the world, there is ample room for a positive Earth emotion based on rarity and uniqueness. The more the uniqueness is understood in, for example, Australia, as a unique assemblage of plants, fungi, digging marsupials, and soil, the more it can be appreciated.

An example of the importance of endemic complexity and diversity can be found in the South West region of Western Australia. As highlighted above, this area has one of the highest levels of plant endemism of any bioregion in the world, and it is a designated world biodiversity hotspot.

Within the region, there is a richness of vascular plants that has attracted international scientific attention. It has 75 percent endemic species, and the numbers of species (over eight thousand recorded) have evolved in the context of highly impoverished soils (low in nutrients) and high background salt levels. In response to these harsh boundary conditions, the WA flora has evolved complex adaptations. In a new understanding that has challenged the old competitive "survival of the fittest" version of evolution, we see, instead, enormously complex symbiotic associations of flowering plants, macrofungi (mycorrhiza), insects, birds, and marsupials all productive of system health and diversity.

In certain areas, the combined outcome of the normal relationship (maintained for millions of years) between marsupials such as Woylies (small, ground-digging rat kangaroos), fungi, soil, and trees is a healthy woodland ecosystem. In woodland sites where Woylies were not to be found, researchers observed that the soil was hard and water repellent due to a buildup of waxy layers from eucalyptus residues that formed a hard surface crust.[102] The water-resistant surface caused high levels of surface water runoff after rain, with major erosion and nutrient transport issues into the waterways. Where Woylies were active, they were able to show that their digging activity broke through the surface of the soil, allowing water and nutrients to penetrate deep into the subsoil, making them available to the whole woodland ecosystem

At the same time, the Woylies were vital to the dispersal of mycorrhizal fungi (truffles) by eating them, then excreting spores back into the soil. The fungi help the nutrient load of the forest. Given that each Woylie can dig and move up to six tons of soil annually, they are a major factor in the overall health of the ecosystem.[103] Such an ecosystem is one in which the maximum possible diversity is maintained by the mosaic of dynamic symbiotic interrelationships between all the elements of the ecosystem. Combined, and including the Indigenous Noongar people, all these elements constitute a unique endemic landscape that cannot be replicated anywhere else on Earth.

Endemophilia captures, in one word, the particular love of the locally and regionally distinctive, manifest in the people of that place. It is what gives a particular sense of belonging, an endemic sense of place, as opposed to a global sense of place. It is similar to what Edward Relph called "existential insideness," or the deep, satisfying feeling of being truly at home with one's distinctive place and culture.[104] Such as state could also be called "homewellness." Relph explains: "The person who has no place with which he identifies is in effect homeless, without roots. But someone who does experience a place from the attitude of existential insideness is part of that place and it is part of him."[105]

In many respects, endemophilia is a precondition for nostalgia, as a person must have an endemophilic emotional attachment to place in order for separation from it to be intensely negative. Likewise for solastalgia, as the chronic desolation of a place is most likely to affect those who love their place and its distinctive characteristics. The more one feels endemophilia, the more likely it is that they intimately know their place, and that intimacy is both a source of terranascient joy and a possible source of negative psychoterratic emotions when terraphthoran forces prevail. If terraphthoran global market forces homogenize landscapes and architecture, the end of endemism appears a likely result. If we move into the Symbiocene, however, such an outcome becomes increasingly unlikely.

Endemophilia opens us to the possibility of a huge array of positive psychoterratic experiences with the flora and fauna of a region. Knowledge of uniqueness plus local phenology enable the same positive emotional experiences from one year to the next. Each year, at the same time, when the orchid species known nowhere else in the world appear in "your place," there is a celebration of that unique event, for it is a sign that all is well with the world. Aboriginal Australians have long enjoyed endemophilia. A translation of a song from the Oenpelli region in Arnhem Land, Northern Australia, captures this special experience:

> Come with me to the point and we'll look at the country,
> We'll look across at the rocks,
> Look, rain is coming!
> It falls on my sweetheart.[106]

In Western Australia, within literary and academic contexts, there has been much done to celebrate and document the uniqueness present in the landscapes, flora, and fauna. I am not alone in seeing the value of the flora and fauna of the South West, and the creative writer and poet Annamaria Weldon, in her book *The Lake's Apprentice*, has expressed her endemophilia for a special region within South West Western Australia.[107] Gradually, as the value of the distinctiveness of places is realized, endemophilia is receiving recognition as a useful concept.

Eutierria

I often drift into unconscious states such as daydreams and lose track of time. Bird-watching does it for me, and I can be "lost with the birds" for hours on end. When I return to consciousness, I remember that I was in a strange state, one that had me totally absorbed in the act of watching other beings and entering in their life forces. In religious traditions, such a state could be described as a form of ecstasy or euphoria, yet I do not feel as if I am in a hyperreal state, nor one that is enhanced by mind-altering substances. However, I am open to the idea that such a state of mind is "spiritual" in some sense, a theme that will be examined in the next chapter.

The feeling of total harmony with our place, and the naive loss of ego (merging subject and object) we often felt as children, have become rare in this period of what Richard Louv calls "nature deficit disorder" (see chapter 3). I realized that such a state of harmony, especially when connected to the Earth, is another emotion that is strangely missing in the lexicon of the English language.

I created "eutierria" to rectify this situation, and it is a positive feeling of oneness with the Earth and its life forces, where the boundaries between self and the rest of nature are obliterated, and a deep sense of peace and connectedness pervades consciousness.[108] Eutierria is derived from *eu* meaning good, *tierra* for the Earth, and *ia,* a suffix for member of a group of positive psychoterratic conditions, in this case, emotion or feelings.

With "eutierria," I have put back into our language, in a single concept, an earthly equivalent of "that oceanic feeling" (connected to religious feelings and / or Freud's psychoanalytic theory) or a secular spiritual feeling of oneness with our home. I feel that Alexander von Humboldt must also have had such a eutierria experience on his travels through Latin America. He described his experience thus: "Nature can be so soothing to the tormented mind, a blue sky, the glittering surface of lake water, the green foliage of trees may be your solace. In such company it is even possible to forget the reality of one's personal existence. It lends wings to our feelings and thoughts."[109]

A NOCTURNE TO EUTIERRIA

In the incandescent glare of artificial light, the deafening roar of the anthrophony and the suffocation of earth smells, a universe of nocturnal experience is lost to us. We are blinded by the light, the sounds of life are silenced by decibels, and the olfactory has been shut down. A whole cosmos of experience is sequestered from our eyes by a haze of glowing tungsten, cool neon, hot halogen, and dazzling diodes that blind us with light. If it is not the cacophony of air conditioners, road and air traffic, it is the digital roar inside the earphones, canceling out the symphony of nature. Empty Nightjars. Dogs pity us when rich, earthly smells are lost as the stench of deodorants gets up our nostrils. To see the night, hear its sounds, and smell its perfume excretions, we must avoid the knock-out lights, clear our blocked noses, and enter a place that is lit by the moon, given voice by ecophony and smell from the olfactory forest. The feeling of earthly oneness, the obliteration of the separation of the self from the other, is called "eutierria" and is "a good Earth feeling." Within a state of eutierria the night is no longer a place of starless black holes, but a good darkness where the senses attune to the harmony of sights, sounds, and smells of our earth and its universe.

We have only just commenced the acts of creation needed to give full expression to our positive Earth emotions. Instead of taking them for granted, as nameless, but good feelings, we now have a typology on the positive side of the psychoterratic to work from. There is still much more work to do in creating the terms for positive Earth emotions. For example, I have thought about "topopinia" as a word for that feeling of wanting to go to places that, in my imagination, I already love and "pine," or long, to go to. For example, I have topopinia for the Highlands of Papua New Guinea and their birds of paradise. I challenge readers to create their own terms for positive psychoterratic Earth emotions in order to adequately describe their own good Earth experiences.

Moreover, it troubles me that we do not have a concept for the opposite of ecocide. I tentatively propose "terraliben" from *terra* (Earth) and *liben* (Old High German, *lebēn*, to live, from Germanic *libēn*)) meaning "let the Earth live." Terranascient thinking might be enough, yet I feel that a strong, simple counter to ecocide and tierracide is urgently needed. I want to be able to punch the air in triumph, yelling "terraliben" with a thousand others, as the terraphthorans either join us or turn to run away. There can be much more, and to get old positive Earth emotions back, and to define and create new ones, will be easier to do now that we have the meme of the Symbiocene as our guide. It is the ideal home in which our positive emotions can thrive.

Stonehenge in Wiltshire, England.

Photograph by the author.

Chapter 5

Gaia and the Ghedeist

Secular Spirituality

I am not a spiritual person in the religious sense, yet as I wander Wallaby Farm and other locations where I can get lost, I feel an affinity with the Earth and its own emotional gravity. I tread where the Indigenous people walked for tens of thousands of years. I can sense their presence as I walk over rock moss that absorbs my bodily vibrations. I feel as if I am being watched and my actions judged. My feelings of eutierria are as close to spiritual, in the sense of pertaining to the nonmaterial, as I ever get, as they involve my complete immersion into a landscape, a place where I am at one with all else. I take photographs of creatures that reside in or are passing through the Wallaby, and I try to focus my image on their eyes. When I see an "alien" eye through the camera obscura, I know that I am being looked at by another being. That look penetrates into my being, and I sense my sensual and mental connection with its owner. We are united in the same place, sharing life for a moment, perhaps contesting it the next (if it is a poisonous snake). When the fireflies are about at dark dusk on the Wallaby, I move into a space called magical realism in literary circles. There is no magic, but in the blink of a firefly, there is something eternal. Under the Milky Way, I feel surrounded by a trillion fireflies and I sense the pattern, movement, and beauty of the cosmos of which I am a part. There is something spiritual in these feelings, and it is a secular feeling, not one connected to God or gods.

In previous chapters I have documented how humans have disengaged from nature and natural forces. In the "great separation" we lose our emotional compass with respect to the Earth and the cosmos. As mentioned in chapter 2, it was Elyne Mitchell who first alerted me to the vital connection between the state of the Earth and emotional-mental states. Psychic instability and solastalgia were allied concepts. She was influenced by Carl Jung, who, while famous as a psychologist and a student of subjective experiences, was also a scientific realist, in that he saw all human emotions and thoughts as issuing from "the universal soil" shared by all other organisms. As a result, when humans become isolated from their "roots," emotional alienation occurs. Jung clearly articulated the loss of a primordial identity with "nature" as a source of the loss of emotional energy in humans:

> As scientific understanding has grown, so our world has become dehumanised. Man feels himself isolated in the cosmos, because he is no longer involved in nature and has lost his emotional "unconscious identity" with natural phenomena. These have slowly lost their symbolic implications. Thunder is no longer the voice of an angry god, nor is lightning his avenging missile. No river contains a spirit, no tree is the life principle of a man, no snake the embodiment of wisdom, no mountain cave the home of a great demon. No voices now speak to man from stones, plants, and animals, nor does he speak to them believing they can hear. His contact with nature has gone, and with it has gone the profound emotional energy that this symbolic connection supplied.[1]

Jung's analysis of emotional loss fits the account of Australian Aboriginal loss under colonial invasion. We saw in chapter 2 that once land-human physical and emotional connections are severed, nothing but turmoil and dis-integration occurs. I have also argued that the emotional upheaval of Aboriginal people is now being replicated in nonindigenous contexts. New forms of invasive colonization, such as large-scale coal mining, have aggressively entered the lives of people in places like the Hunter Valley and Appalachia. This process has now gone global under the relentless impacts of climate warming. We all now live in a small village called Bulga that is being desolated by coal mining at the same time as it gets hotter.

Elyne Mitchell thought it possible to restore our involvement with nature and recover our emotional or spiritual health. Mitchell concluded her prescient book with an appeal to her readers to build a symbiotic civilization:

The rhythm of night and day, of seasonal change, of growth and decay, is a living pulse with which our own lives must beat in harmony—the constant change enfolded in our life span more obvious than the trees "yet slower than the lambs." And the perpetual organic return ensures that the cycle may be continued in the darkness of the soil—through the current of life that is never broken except by man. Here is the living expression of the interdependence and unity that says "all things are one," the living expression of immortal fire and the fire of generation; here the flowing river that is never the same yet unchanging.[2]

In her nonfiction publications, Mitchell emphasized that her foundation for a new symbiosis between humans and the land was ultimately dependent on what she variously called a land *Geist*, land spirit, *spiritus mundi,* and the Vital Spirit.[3] To feel such land- or earth-spirit affinity, Mitchell also thought that the "strange rhythm and harmony that is love" is not isolated within individual humans but is a fractal of love within the wider cosmos. I will reengage with the concept of love below.

In what follows, I will replace the past Jungian symbolic connections to Earth emotions with new symbiotic connections. I will endeavor to see if there is an emotional "unconscious" within the symbioment, which can be lifted into consciousness, and explore that consciousness as a foundation of the Symbiocene. There is also love, and the idea that one of the most intense emotions we still have within us as human beings is the capacity to receive and give love. To take love from the domain of individual humans and project it out into the realm of the Earth, as alive, as a "person," is an idea shared by many. Can we love the Earth and, in doing so, form bonds so tight and strong that the achievement of the Symbiocene will be a walk in the park? Given that the Symbiocene is such a revolutionary idea, it is going to take some radical thinking to help it take hold in the garden of ideas.

Gaia and the Anima Mundi

In 1988 I completed my PhD thesis in philosophy on the topic of organicism. In the thesis I argued that there was a tradition in Western philosophy that gave emphasis to the organic unity of life in general and human social life in particular. The idea of organic life, its properties and characteristics, was used as a metaphor and conceptual foundation for all other forms of complexity within human society and its institutions. I argued that the pre-Socratics, the classical Greek philosophers, and especially the German idealists such as G. W. F. Hegel used an active, purposive, and organic view of life to create a systematic organicist philosophy, and an ethic attached to that view. As much as I would like to give you a summary of the whole thesis, including the detail of Hegel's dialectic as a philosophical expression of the vitality of life, it was the connection I made to the environmental ethics of the late twentieth century that is relevant here. I argued:

> It is somewhat ironic, although conducive to the organic unity of this thesis, that one of the most influential theories propounding the interrelatedness of all life in recent times has been Lovelock's Gaia hypothesis. The theory that the Biosphere is or acts like a *living organism* returns us to an idea that was present in pre-Socratic thought. The idea that organic unity and order pervades the whole biosphere and perhaps even greater domains of the reality continuum brings with it the recognition that the ethical or the good is a "part" of the objective order of things.[4]

The idea that the planet is organically unified can be extended, on the basis of the new science of symbiosis, into the realization that the Earth is the ultimate symbiotic biome. Life on Earth, as symbiotically unified, has strong evidence supporting it at all levels of the biosphere.

While the evidence for a symbiotic biosphere is now strong, I see no such evidence as yet for a symbiotic, living cosmos. There has been no sound evidence presented of life existing outside the boundaries of the Earth. I am not even convinced by the idea of Gaia as a self-regulating superorganism, animated by symbiosis at a planetary scale.[5] There is symbiosis in and between life forms, but not as a factor driving life as "Gaia," or the total Earth. It is for this reason that I share with Lynn Margulis a reservation about Gaia being "alive" and having the characteristics of personhood.[6] However, I recognize that others, inspired by the idea of Gaia, see it as having "spiritual" and/or "personal" characteristics. For those who see the Earth as alive, the Gaian worldview is more than a materialist explanation of the way the world works; its animism also gives us a spiritual dimension to life.[7]

One concrete application of the Gaia hypothesis is in the domain of ethics, particularly for our relationships with nonhuman beings. Following in Aldo Leopold's footsteps, and his idea of developing a sense of "kinship with fellow creatures," a Gaian-inspired ethic can be built on the Darwinian connection of species evolving from a common ancestor over time, plus an ecologically and symbiotically grounded nexus with other beings in a shared space. Stephen Clark, in the essay "Gaia and the Forms of Life," published in 1983, took the now popular idea of kinship with other species into the realm of formal environmental ethics. He argued:

> The awakening of this sense of kinship, this making real to oneself that we exist, and can only exist, as elements within modes of a continuing community (not an aggregate or socially contracting multitude of separable individuals) is the realistic analogue of the Kantian "kingdom of ends." Its ethical imperative might be represented as follows: "So act as if your maxims had to serve at the same time as universal law for all entities that make up the world."[8]

Crispin Tickell, in the foreword to James Lovelock's *The Revenge of Gaia*, quotes from Lovelock on symbiosis as "a lasting relationship of mutual benefit to the host and the invader." He writes that Lovelock maintains that the only viable future is to have symbiosis, not a disease relationship, with the rest of the planet.[9] Tickell then asks the question: "How do we achieve that symbiosis?"[10]

I engage with this important question by offering a response based on the role of our emotions. The role of positive Earth emotions will be reinforced by putting them all within the context of a secular land or Earth "spirituality." Rather than rely on a spiritual interpretation of Gaia, or appropriate spiritual affinities to the Earth displayed by indigenous peoples, I put the case for a newly formulated life spirituality based on symbiotic science that is relevant to the twenty-first century and beyond. Positive Earth emotions will triumph over the negative, and will do so by establishing the love of the Symbiocene, animated by a secular form of spirituality.

Avatar and Sumbiocriticism

Before I explore some select literature on love and spirit, I want to examine the 2009 film *Avatar* by James Cameron from the perspective of sumbiocriticism.[11] I apply sumbiocentric critical evaluation to the major themes of the film. My use of *Avatar* is not for its filmic qualities, but for what it portrays as a fictional Symbiocene setting with themes that are relevant to the major concerns of this book, especially the Anthropocene and its colonizing tendencies. I wish to disconnect myself and my evaluation from *Avatar*'s filmic qualities.

Sumbiocriticism, as I have defined it, is a branch of ecocriticism that evaluates all forms of creative endeavor from the perspective of their awareness of

- the degree of interconnectedness between the social world and the biological and ecological systems that support it;
- the extent of the symbiotic interconnectedness between different types of beings on this planet;
- the ability to convey the love of a community of beings living together on this Earth (sense of kinship); and
- the contribution to the idea of sumbiosic development and the goal of living together within the symbioment.

The setting for *Avatar* is the moon Pandora, where a tribe of human-like species, the Na'vi, are resident. A rapacious group of humans invades Pandora seeking "unobtanium," a mineral resource that is rare and hugely valuable. The Na'vi are a barrier to the free access to unobtanium, and much of the plot of the film revolves around the conflict between humans and the Na'vi. A diplomatic link between the humans and the Na'vi is created in the shape of a genetically formed "hybrid" with the mind of a human and the body of a Na'vi. The hybrid, or "avatar," is able to communicate with the Na'vi and interact with them in their own habitat.[12]

The life on the moon Pandora is organized around a combination of cooperative mutualism, a deep ecological vision, and symbiosis on a grand scale. There are examples, within the fauna of Pandora, of organisms that have mutually beneficial symbiotic relationships with each other. But it is the grand-scale symbiotic mutualism that comes closest to the current scientific understanding of symbiosis, within the "wood-wide-web." On an *Avatar* website it is suggested that "the trees and plant life of Pandora have formed electrochemical connections between their roots and effectively act as neurons, creating a moon-wide 'brain' that has achieved sentience, which is known to the Na'vi as Eywa."[13] And, to build the plausibility of the fiction further:

> From what scientists can tell, the Pandora ecology works and communicates like a nervous system, suggesting a symbiotic relationship between all things Pandora. Perhaps the best symbols of this relationship are small luminescent Wood sprites, which are the seeds of the d'Utraya Mokri, the tree of souls. The tree is sacred to the Na'vi and believed to be the heart of the deep connection of all life.[14]

· Not only does Eywa build life, it also can return life to the dead. It connects all life forms together so that they can cooperate and ultimately battle the human invaders.

There is one scene in the film that I want to focus on. The human male, Jake Sully (in avatar Na'vi physical form) and the native female Na'vi Pandoran, Neytiri, intertwine their prehensile, braided ponytails, or "queues." Because these queues are extensions of their neural systems, they can be used for communication, collaboration, and connection with other Na'vi, different species, and the totality of life on Pandora. When Jake and Neytiri conjoin the sensitive tendrils at the tip of their queues, there is a visible erotic electricity, and their union is seen and felt to be sensual (possibly sexual) and intensely cognitive. Their symbiotic moment could also be a manifestation of what could be called "love" in sentient, sensitive beings. It can be argued that to fully unite with the other in a shared life is the ultimate expression of love. It is, as Hegel would call it, the fulfillment of "identity-in-difference." I think it is more eloquently expressed as the "symbiotic erotic." Jake and Neytiri, by conjoining, allow the audience to develop deep empathy with the Na'vi.

The film can easily be read as an allegory on the plight of the contemporary Earth. In their rapacious and murderous quest for "unobtanium." the humans who invade Pandora care nothing for the distinctive humanoid and other forms of life. They have no real understanding of how the Na'vi and other life forms on Pandora are interconnected. They engage in a great military drive for the extinction of any and everything that gets in their way. They are willing to use powerful technologies to blow up and burn the "tree of life," the soul tree, in order to get more material wealth and power. Humans, for the most part, are portrayed as a species with no empathy for the greater forces that hold life together, and put on full display their willingness to commit war and ecocide to achieve their aims. A very bleak view of human nature is contrasted with the alien, animistic, pantheistic, and ecologically "spiritual" humanoids. It is a battle between terraphthora and terranascia.

When the film was playing in cinemas, it was observed that many in the audience wanted to go back and recapture the intense life and love experience of the Na'vi for each other and their Gaian "living moon," Pandora, immediately after the film concluded. Many reported experiencing "the blues," or what I interpret as powerful, negative psychoterratic emotions in the form of both intense nostalgia, the desire to return "home," and what I have called a type of virtual solastalgia, or deep existential distress about the environmental desolation that happened on Pandora at the hands of the human invaders. People were emotionally desolated on "coming back to Earth" at the film's conclusion, despite the fact that Pandora only existed in the cinema as an enhanced 3-D glasses experience. One viewer succinctly described these emotions: "One can say my depression was twofold: I was depressed because I really wanted to live in Pandora, which seemed like such a perfect place, but I was also depressed and disgusted with the sight of our world, what we have done to Earth. I so much wanted to escape reality."[15]

I wrote about the psychoterratic dimensions of *Avatar* on my blog, Healthearth, not long after viewing the film in 2010:

> As the real world is being desolated (climate change, ecosystem distress, etc., etc.), real people experience solastalgia. When, in *Avatar*, they can "see" an alternative world, which is beautiful, diverse and complex, one that meets their aesthetic, spiritual and ethical needs, they want to live within it. During the three dimensional movie, they experience a virtual solastalgia as they become virtual participants in the attempted destruction and desolation of the Na'vi and other life forms in this pristine environment . . . all for the sake of a meaningless materialism. The movie becomes, for such people, an existential experience of negative environmental change (defined as solastalgia). At the conclusion of the movie, when they must accept that such a world is virtual only, they experience a virtual nostalgia for it and become depressed. The irony of humans finally seeing the value of life, different ways of being "human" plus the intrinsically valuable complexity of non-human beings and their living systems via a movie about a virtual world and its destruction is not lost on me.[16]

What the *Avatar* virtual solastalgia experience suggested to me was that deep inside many people was the empathy needed to commune with the coherent and complex world that planet Earth once represented for all humans and all life. Not necessarily a perfectly peaceful and harmonious place, but one that sufficiently favored the forces of creation over the forces of destruction, to allow life and the positive emotions to be nurtured and perpetuated. Pandora was a fantasy of a place that allowed expression of psychoterratic emotions such as solastalgia, nostalgia, soliphilia, and eutierria. In addition, mutualism via symbiotic "unity-in-diversity" enabled the forces of life to finally overcome the forces of death and destruction. I am not the first to notice the connections between Earth emotions and Pandora emotions, and the psychoterratic distress felt by people as a result of their engagement with the movie.[17]

People who remain connected to the Earth in their everyday lives have not lost this emotional empathy. But perhaps intensely urban people can only now experience this kind of primordial empathy for other life forms and the systems that sustain them when they experience it in a virtual setting such as a film. It suggests that there is something foundational and fundamental in the psyches of people that leads to the rejection of the deliberate destruction of life, and that affirms a love of life and a life of love. As documented in the previous chapter, I have named this affirmation of life "sumbiophilia," and it sits alongside biophilia in the positive psychoterratic typology.

Since the first showing of *Avatar*, the state of the world, by all forms of symbiomental measure, has only worsened. Powerful corporations are still searching the world to extract and burn fossil fuels. Places where indigenous people live traditional lives continue to be "developed" and exploited for yet more fossil-fueled fire. As the Arctic melts, the fossil fuel industry sees it as an opportunity to extract more oil, rather than an unmitigated disaster. Climate change is causing whole regions of the planet to become uninhabitable, to the point where all forms of life have to "relocate" to safer places. Even those in so-called developed countries are now finding their health status, life span, life chances, and quality of life to be seriously inferior to that of their own parents' generation.

As the world goes to hell in a handbasket, I wonder if there is enough love left in the queues of humanity to save us from the forces that drive death and destruction. What relics of a collective love- and life-affirming spirit within humanity are there to counter the ecocide, xenophobia, and collective violence emerging in many countries and their leaders worldwide? If it is not evident as a mass movement in the real world, then it was certainly evident in some of the many people who viewed *Avatar*.

Avatar was clearly a work of science fiction. Yet the advance of the science behind symbiosis at the microscopic level has enhanced the credibility and veracity of the film's vision of life on Pandora. The film enables people to see a humanoid culture strongly connected to its symbioment and other holobionts sharing life. Viewers were also able to see how people could communicate and connect in nondestructive ways, and how future generations of Na'vi were to be nurtured by both society and the tree of life.

The kinship, or sumbiophilia, with other species identified via Leopold and the Gaia hypothesis also finds resonance in the relationships between Na'vi and non-Na'vi explored in the film. The love of life and the respect for life can also be transferred between human beings in the uniting power of love. The tree of life appears as a spiritual unifying force, where the power of unity and creation is stronger than the power of destruction. The good humans, who join the Na'vi in defeating their fellow human terraphthoric forces, are judged as heroes. The film also demonstrated that there are many ways that a collective sense of nonmaterial or spiritual empathy for life and its cooperative side can be expressed, including the cooperation required to fight battles and win out over evil.[18]

Finally, to complete the sumbiocriticism of the film, the Na'vi and the organic unity displayed on Pandora display all the characteristics of the proto-Symbiocene as outlined in chapter 4. *Avatar* was a film that presented the proto-Symbiocene at work in a fictional setting.[19]

All this takes us back to Don Fredericksen and his Jungian warning that as we view and comprehend ourselves as vicariously complicit in the attempted destruction of Pandora, we must confront our own personal terraphthoran tendencies. *Avatar* is a movie where we watch our own species attempt to destroy many others. We are forced to watch the terraphthoric within us that we normally would evade. No wonder people felt an extremely uncomfortable form of virtual solastalgia at the attempted destruction of Pandora and the tree of life. They felt it because they know they are participants in the same archetypal drama happening here on Earth. To want to go back to Pandora in its pristine state is a form of evasion. It is worth quoting Don Fredericksen once again on this point: "But knowing that I know that I do not want to acknowledge yields the most painful awareness. I mean simply this: by evading my own titanic psychological register, evading it perhaps titanically, I am polluting my own psychic home. This generates an inner, self-generated form of solastalgia."[20]

To want to return to Pandora is a nostalgic solution to this inner solastalgia, but it is also a form of identity with those forces that affirm the goodness and beauty of life. The union between people and place, the union between people and other beings, and the union between people and other people are all present in Pandora to help us see what we need to do on Earth. *Avatar* is a massive love story about the union between two individuals, a form of erotic symbiosis, and the union between people and place, ecological and cultural symbiosis. It engages the human emotions at every possible level, including Earth emotions, so no wonder it was such a success—as its prequels and sequels will be if they are ever made.

It is not fanciful, then, to think about the symbiotic connections between species, and the intense collaboration evident within species, as a kind of empathetic life "spirit" that holds it all together. Perhaps it was a "whiff" of this spirit that people watching *Avatar* felt so strongly. Moreover, we should not be all that surprised at such spirit recognition, because the symbiotic empathy in the Na'vi and their tree of life has antecedents in all human culture, including in Eurocentric mythology. Elyne Mitchell mentions "Yggdrasil," the giant ash tree in Norse mythology.[21] It is the tree of life; it connects all living things via its leaves, branches, and roots and nurtures life through its cycles of life and death. The tree of life symbolism can be found in a great many human cultures. Clearly, James Cameron was tapping into something far deeper than modern symbiotic science as he developed the plot of *Avatar*.

The Love of Small Things

The sense that there is something larger in life than an individual has seen humans look to the largest living organisms on Earth for inspiration on what the meaning of life is. Trees are an obvious example of visible unity in diversity, a prime example of what the philosopher Hegel called the dialectic, where the distinctive, temporally distant parts of the plant structure—bud, blossom, and fruit—all cohere in a single collaborative entity. They are all "moments" of the one living entity "in process" over time. Trees can also be inspirational, as explained by John Christie, a writer who runs a heritage farm in Australia. He examines the idea of a primordial connection to trees and their connection to empathy:

> In mythology, this is the World tree, the "axus [*sic*] mundi," the centre uniting the heavens above with the underworld below, the still point around which the universe revolves. Here the tree is a centring force. Do we find comfort here? Robert Vischer was a German philosopher who invented the term "Einfuhlung" (aesthetic sensibility), later translated into English as "empathy," in the late 1800s.
>
> This idea developed from an older concept that linked thinking and bodily feeling with an appreciation of objects. For example, it is said that a body swells when it enters a hall; it sways, even in imagination, when it sees wind blowing in a tree. It is the experience of the rhythmic continuity between self and other, between "outside" and "in." The suggestion here is that we objectify the self in external, spatial forms (in this case trees), projecting it into and becoming analogous with them, merging subject with object.

We become the tree; self and world unite. Trees, or groups of trees, then, are like botanical cathedrals—cavernous and protective and uplifting—and it is to this that we respond in an "empathetic" way. Perhaps.[22]

I agree with Christie, in that my concept of eutierria, or a "good earth feeling," also tries to capture that sense of the merging of subject and object. I was thinking of smaller units of place than the whole world, but why not? If we can empathize with the tree of life, merge with it, then we could be in a perpetual state of eutierria.

However, we have been barking up the wrong tree when it comes to adoration of size. The largest living organisms with a contiguous genetic identity are not giant trees or whales; they are colossal networks of fungi. Indeed, in 2007, it was reported that

the discovery of this giant Armillaria ostoyae in 1998 heralded a new record holder for the title of the world's largest known organism, believed by most to be the 110-foot- (33.5-meter-) long, 200-ton blue whale. Based on its current growth rate, the fungus is estimated to be 2,400 years old but could be as ancient as 8,650 years, which would earn it a place among the oldest living organisms as well.[23]

While an interesting news story, this account of the "biggest" (and oldest) fails to really appreciate the fact that the fungus is embedded within an even larger organism, the forest, and its whole array of symbiont biomes. Moreover, as we have observed, within the fungus are various types of symbiont bacteria that help the fungus survive invasive bacterial attack. This collective entity, as we now understand through plant science, can interconnect trees in a forest, and via the symbiotic fungus mycorrhizae, extend the root network over hundreds of square kilometers, and perhaps even farther, when we consider the myriad other symbiotic connections between organisms.[24] Perhaps we were also barking up the wrong tree when considering size as the marker of life. While it is easy to love a "big," powerful god, a huge tree, or even an extensive forest, it is difficult, if not impossible, to love symbiotic bacteria within one's gut and in every other orifice and surface that we have on our bodies.

The great thinkers of the past, when seeking inspiration, could not possibly include in their ratiocinations the idea that at the personal bodily level, the greatest number of cells, trillions of them, that constitute active, living parts of the human body—roughly 90 percent of them are in fact symbiont bacteria. As Bassler reminds us: "You have 10^{10} bacterial cells in your gut.

You only have 10^9 human cells making up your whole body. So there are 10 times more bacterial cells in you, or on you, than human cells. By weight, you are more human than bacteria, because your cells are bigger, but by numbers, it's not even close."[25] We can add, to the single-celled bacteria, a good number of fungi and viruses as well. We have yet to discover the sum total of life forms that live with us in the human holobiont, or the host house we call the human body.

Nor could the great thinkers have guessed that, invisible to the eye, bacteria are by far the largest family unit in the great tree of life.[26] If bacteria and mycorrhizae had been visible to us, Descartes and other founding fathers of mechanistic reductionism would have immediately seen the need for holistic life sciences. We are only just beginning to appreciate that our life in the symbioment is a shared arrangement, a mutually beneficial relationship between holobionts, radically different life forms that share life and death in common. All creatures great and small will work together to keep their share of life for as long as possible.

We are even discovering new places where the great diversity of life can be found. New research has discovered that, within the Earth, there are novel symbiotic microbiomes. As Gaetan Borgonie and Maggie Lau reveal:

> A widespread misconception about the deep subsurface is that this realm consists of a continuous mass of uniform compressed solid rock. Few are aware that this mass of rock is heavily fractured, and water runs in many of these fractures and faults, down to depths of many kilometres. The deep Earth supports an entire biosphere, largely cut off from the surface world, and is still only beginning to be explored and understood. . . .
>
> More interesting, we deduce that deep microbial groups have established strong, paired metabolic partnerships, or syntrophic relationships, which help the organisms overcome the challenges of extracting the limited energy that originated from rocks. Rather than competing directly with each other, these microbes establish a win-win collaboration.[27]

Now that we know there is a deep Earth microbiome, it suggests that we must take extra care when drilling into the Earth. Pushing water and chemicals into the Earth in search of gas (fracking) and the subsequent release of huge amounts of wastewater involve major disturbance of this realm of life. Where is the section, in the current environmental impact statements for fracking, on the impact on deep symbiotic microbes? Breaking symbiotic bonds is beyond an "impact": it is an assault on a deep foundation for life.

The next big question is how we should view the symbiotic arrangement in life in general, and human life in particular, in metaphysical terms. For a start, we have to think about our identity. We are not who or what we thought we were. We were very comfortable with the idea that we were autonomous entities and highly evolved "individuals." Quite rightly, we should now feel discombobulated!

Some humans have managed fairly easily to conceptualize being part of the "grand" half of the symbiotic turn in thinking about life. It has been both familiar and comfortable to think that we are part of something much larger than ourselves (Gaia) as such a view is consistent with older traditions of thought. However, the other half, if I can put it that way, is to think about what it means to be a vital "part" of something much smaller than ourselves (microbiomes). Our identity and life course is as much tied to microbes and microbiomes as it is to macrobiomes (ecosystems). Our empathy for life must expand, both in the presence of big trees and the presence of trillions of microbes alive in and on us. Symbiosis and empathy, at small scales, should not be strange bedfellows.

Love as Invisible Indivisible Unity

Beyond empathy toward specific entities, there is the idea that, pervading whole complex ecosystems, there is a unifying force or vital characteristic that is shared by all within it. Even here, we abstract too much, as ecosystems are a human fabrication of system boundaries that do not exist in the symbioment. Life in the symbioment knows no boundaries or walls.[28]

An early thinker who took on the task of understanding life within a *boundless* ecologically connected entity was Rudolf Steiner. In 1923, Steiner created the concept of "love life" in relation to the life of bees. He argued:

> That which we experience within ourselves only at a time when our hearts develop love is actually the very same thing that is present as a substance in the entire beehive. The whole beehive is permeated with life based on love. In many ways the bees renounce love, and thereby this love develops within the entire beehive. You'll begin to understand the life of bees once you're clear about the fact that the bee lives as if it were in an atmosphere pervaded thoroughly by love . . . the bee sucks its nourishment, which it makes into honey, from the parts of a plant that are steeped in love life. And the bee, if you could express it this way, brings love life from the flowers into the beehive. So you'll come to the conclusion that you need to study the life of bees from the standpoint of the soul.[29]

Studying bees from the standpoint of the "soul" can be roughly translated as studying bees in the context of their relationship to flowering plants and, in turn, plant relationships to fungi, or what Elyne Mitchell would call the earth-spirit of the soil. The idea that bees capture "love life" and turn it into honey is a wonderful way of describing something within science that is very complex. I could tell a story of enzymes, sugars, a bee honey stomach, microbiomes within that stomach, regurgitation, and evaporation, but that simply avoids the fact that bees convert life in plants derived via symbiosis with fungi and their own symbiont bacteria into honey. Honey is a holobiont's product of life; there is more to honey than bees.

Steiner was a pioneer in the study of organic interconnections and saw life and love in ecological terms. He suggested in his 1923 lectures on bees that to understand life, "you need to take a deep look into the entire ecology that nature has to offer."[30]

In making the case for "love life," Steiner was following many others in the history of thought who saw affinities between love and life. I suggested earlier that we could learn a great deal from indigenous people about what an Earth-based spirituality consists of. The Dreaming, for me, was and still is a form of symbiotic spirituality that unites people to the symbioment and the cosmos. To see your "country" as your "sweetheart" means that you love it. In addition, although not able to see or directly experience symbiosis at the microscopic scale, traditional Aboriginal people accepted that you do not need to see everything to know that all is interconnected. The late Big Bill Neidjie, in his educative account of the traditional beliefs of his people in Arnhem Land, published in *Gagudju Man,* makes it clear that unknown forces are part of what he calls "sacred places." Because I viewed a documentary on this project, to this day I can still hear his deep, earthly, resonant voice inside my head saying:

> We walk on earth,
> We look after,
> like rainbow sitting on top.
> But something underneath,
> under the ground.
> We don't know.
> You don't know.[31]

Despite the fact that we are on the path to "knowing" a lot about the invisible and the subterranean, Big Bill's statement still holds true. We know the component parts of the invisible networks that cohere as a shared life, but we are not yet able to define life itself, except in terms of a shared life "property," the biocomunen, between disparate elements of an organic whole. Perhaps there is nothing new in this insight.

Steiner's idea, that sharing the property of life is a form of love, has been a long-standing theme in Western and non-Western thinking. Other, older cultures have also written about the power of a love spirit or essence that pervades all things. The ancient Greeks, for example, saw life in terms of a struggle between the forces of unity and dissolution. The pre-Socratic lover of wisdom Empedocles wrote about love and strife, and he maintained that only when love "is at the centre of the whirl, in it do all things come together so as to be one only." In Plato's *Dialogues* the need for balance between love and strife is discussed, and in the *Symposium*, Eryximachus maintains that, for health to flourish, "he who is able to separate fair love from foul, or convert one into the other, . . . can reconcile the most hostile elements in the constitution and make them loving friends."[32]

Hegel "in Love"

Nineteenth-century philosophers such as G. W. F. Hegel and Goethe also saw love as foundational to life. Hegel's whole philosophy might be thought of as a philosophy of life, where the blueprints for all kinds of nonmechanical relationships are organic unity in living organisms and the unity we call love between humans. Hegel's fragment "Love," the title given by his editor Herman Nohl, was written between 1797 and 1798. In this early work Hegel develops a formulation of organic, living relationships that will be carried through to his mature writings. He begins the fragment on love by arguing for the reality of relationships: "In fact, nothing is unconditioned; nothing carries the root of its own being in itself (subject and object, man and matter), each is only relatively necessary; the one exists only for the other."[33] In the condition of "love," this relational feature is clearly manifested. Love is, according to Hegel, a true union as it "exists only between living beings who are alike in power and thus in one another's eyes living beings from every point of view; in no respect is either dead for the other."[34]

Hegel sees the possible result of such a union, a child, as having a developmental process to bring it to unity within a loving relationship. In an early formulation of the dialectic, Hegel describes this form of development: "The seed breaks free from its original unity, turns ever more into opposition, and begins to develop. Each stage of its development is a separation, and its aim in each is to regain for itself the full richness of life. . . . Thus the process is: unity, separated opposites, reunion."[35]

Hegel would go on in his later writings to describe this process of love in terms of life, then the Absolute, or spirit. Yet the "process" philosophy underlying them all is an attempt to "think life." Hegel's famous idea of the "dialectic" is the mature formulation of this quest to describe life within a "process" philosophy. Despite later, more complex accounts of the dialectic, much of it using organic metaphor, he produced an account of the dynamic process that animates life in his early writings. Hegel, in unusually clear language, says: "In love man has found himself in another. Since love is a unification of life, it presupposes division, a development of life, a developed many-sidedness of life. The more variegated the manifold in which life is alive, the more places in which it can sense itself, the deeper does love become."[36] That is the kernel of truth about love: it is paradigmatic of a spirit or essence that is both singular and universal at the same time. It is this quality that marks love as a type of organic unity and of great philosophical importance. Hegel, to give him the final word on love, says that it is "something which is at one and the same time feeling, i.e. in the heart, and object; feeling here means a spirit which pervades everything and remains a single essence even if every individual is conscious of his feeling as his own individual feeling."[37]

I will take Hegel's lead in defining love as a spirit that pervades everything, as the basis for a secular spirituality, based on the symbiosis that unites life in all of its uniqueness and diversity.

Love and Symbiosis

From the twentieth century onward, serious arguments about the similarities between love and life have been few and far between. The very idea that one could think of the relationship between person and place, body and symbioment, as similar to one between lovers seems to have no relevance in a world now obsessed and assessed, mainly by value-neutral science, materialism, and utility cash values. The "philias" of life have been either neutralized or commercialized, while philiaphiles, such as artists and writers, struggle to find empathetic audiences for their expression of the "environmental" crisis as a loss of love. It is perhaps much easier for the media to cash in on solastalgia and depression and present them as victims of the ecoapocalypse. Everyone wants to look at a disaster.

As was mentioned in chapter 4, in the 1960s Erich Fromm developed the idea of biophilia, or the love of life. He argued that the biophilic personality wants to "mold and to influence by love, reason, by his example" rather than by authoritarianism and force.[38] The associative nature of love will be discussed below; however, Fromm takes biophilia into every aspect of human development, including the personality. He sees biophilia as a "total orientation" toward living, and says that, for men and women, it "is manifested in a person's bodily processes, in his emotions, in his thoughts, in his gestures; the biophilious orientation expresses itself in the whole man."[39]

The contemporary Welsh writer Ginny Battson has explicitly taken love-life connections between humans and elemental symbiotic forces within biomes into a new dimension. She writes about the fungal hyphae networks and their analogy to the idea of love:

> Hyphae grow from their very "finger" tips, the softest exploration in finding a way to their next interconnection. In a lab, the direction of hyphal growth can be controlled by environmental stimuli, such as the application of an electric field. Hyphae can sense reproductive opportunities from some distance, and grow towards them. Hyphae can weave through a permeable surface to penetrate it.
>
> One may consider the human spirit of love a little like the hyphae, in sensing partners and finding ways to connect and exchange through layers. Love itself, of course, glows in many rainbow colours. Aristotle says love is composed of a single soul inhabiting two bodies. Mycelium may be the soul and unity of the forest, where not just two beings are united, but many, and for the love of the whole community.[40]

In the human context, cooperation between people is a foundation for all forms of social activity. Love, empathy, and friendship are essentially cooperative in nature, and they are the basis of further creative and nurturing acts. Our work as parents, its intrinsically nonselfish nature, helps the child mature into a productive and caring adult. The same process applies to all forms of productive work, including art and industry. Love as cooperative and associative is a form of cultural symbiosis because it cannot be compelled but is freely undertaken and shared to achieve an expanding domain of goods. Love is the foundation for ethics.

The sensing of the needs of others is paradigmatic of love, and it is freely undertaken by those who are "in love." As argued by the Australian philosopher Eugene Kamenka, "The free quality of love lies in the fact that it does not require external restraints or internal illusion and repressions in order to continue as love."[41]

The Ghedeist

We can see that, for life and love, both are only possible when there is connection, exchange, creation, reproduction, defense, and sharing. The principle or biocomunen which animates life in a forest or on the human gut-brain axis is one that has a pattern that is repeated in love between human beings at a small scale (between lovers) and at the larger scale in the soliphilia or "love of the whole community" of humans. With individual organisms, the unity can be shattered by disunifying forces, such as toxic substances, cancerous growth, and rampant infection. Yet for life to endure, the forces that create and perpetuate life must prevail, in the long run, over the destructive forces. For life to be generally self-destructive is a contradiction that would end in extinction. More than an "equal balance" must be struck, as life forces manifest as organisms and must be able to grow to maturity, complete reproductive cycles, and keep destructive forces at bay.

At larger levels of organization, the sciences of biology and ecology have shown us that species are interconnected in ways that maintain the health of whole communities. The property of "ecosystem health" is achieved, as we observed in chapter 4, by the interaction of a diverse number of species all working in a coordinated way to achieve a common end. Ultimately, large-scale organization is reflected in small-scale organization and vice versa. At all levels, mutual cooperation, the symbiosis between diverse species, is an expanding outcome of evolution by symbiogenesis over time.

When diverse organisms and diverse species act together, they share the biocomunen, the common property of life. No one organism by itself can achieve this end. Bacteria, plants, and people all cooperate among themselves and mysteriously "communicate" their needs. Even gut bacteria communicate among themselves, a process called "quorum sensing," to maximize their ability to coordinate and help each other and their symbiont partner (us).[42] Such an insight is also practical in that it helps us understand that repairing the damage to the Earth is a cooperative enterprise between many different types of organisms. Germaine Greer, while restoring ecosystem complexity and health on a degraded property at Cave Creek in south eastern Queensland, had the worrying thought that the task seemed too great for her:

I walked to where Brush-turkeys had scraped all the expensive mulch into a huge new mound inside which their eggs were already incubating. All my anxiety ebbed out of memory. How could I have thought I was in this by myself? I had helpers, thousands, no, millions of them, as well as five humans. Cave Creek wasn't just another anthropogenic biome after all. We were all working together, bacteria, fungi, invertebrates, reptiles, amphibians, birds and trees, plus the odd human.[43]

There remains something deeply mysterious and vague about the way the systems of interconnection work. We have yet to figure out how "they" communicate with "us," and us with them. Yet it happens seamlessly a million times over, every moment of every day within a human body. Sure, there is the vagus nerve, gene signaling, and chemistry, but it goes further than that, as a mystery, the shared biocomunen. That mystery opens space for a new secular concept for the life-affinity between diverse organisms. Strictly speaking, they are not all our "kin" since, in the case of bacteria, we do not even know that they are there. Besides, there are some in the ecosystem I do not wish to share kinship status with. Paralysis ticks at Wallaby Farm, for example. They are blood-sucking parasites that can also kill their host. No mutualism here.[44]

As a result of the mystery of the property of shared life, the idea of a secular word for "spirit" has also exercised my mind. Very few humans have developed a material affinity with the trillions of organisms that are essential to our life but are beyond our perceptual physical and psychological boundaries. Apart from some recently acquired knowledge that our gut microbiome might be vital for physical and mental health, we have yet to incorporate the presence of our symbiotic gut flora into any kind of empathetic or "spiritual" realm. I put the case that the presence of the biocomunen within holobionts forms the foundation for a secular form of spirituality that unites the interconnected life between all life forms and all individual beings on the living planet. We require new concepts to help us have emotional contact with the whole of life, not only that which is big, spectacular, and directly present to our senses.

There is an old root word in Indo-European languages that captures the meaning I want to convey. "Ghehd" has meanings linked to Old English and Germanic words such as "together," "to gather," and "good." I have thought that a modern version of the word "ghehd" could be incorporated into a secular spiritual context in the form of the "ghedeist." The word is formed with the ghost, ghed, and a shortening of the German "*Geist*," with its meanings of spirit and mind and affinities in other languages of a vital force or "life force." The neologism "ghedeist" was thus created by me to account for a secular positive feeling for the unity of life, and the intuition, now backed up by science, that all things are interconnected by the sharing of a life force. My definition of the "ghedeist" is the awareness of the spirit or force that holds things together, a secular feeling of interconnectedness in life between the self and other beings (human and nonhuman) and their gathering together to live within shared Earth places and spaces, including our own bodies. It is a feeling of intense affinity and sense of empathy for other beings that all share a joint life. It is a feeling of deep association with the grand project we call life.

We no longer need "faith" to believe in something of which we normally have no direct sensory awareness because, via microscopic technology, we can actually see symbiosis at work in microbiomes. Education and experience to understand symbiotic spirituality can take place largely through immersion in nature at the large scale, and immersion into the microscopic by the use of sense-aiding technology.

Some of this technology can be digitally enhanced, especially when science fiction films like *Avatar* provide a model of how people can be brought into the presence of symbiotically connected life in three dimensions. An earlier movie, *Fantastic Voyage* (1966), took a team of miniaturized people inside a miniaturized submarine into the damaged brain of a human in order to conduct vital repairs to the neural circuits. I can imagine a three-dimensional educational journey through the gut that takes the viewer from a starting point at the mouth and its oral microbiome, through the alimentary canal, and then exits via the anus. Along the way, we meet the gut flora and the good bacteria that keep us healthy. We also meet the bad bacteria and see how they are defeated or neutralized by the goodies. Could the film *Journey to the Bottom of My Body* be a blockbuster? Authors of children's books could turn the whole symbiotic microbiome story into a new genre.

Ghedeistual Ethics

The associative nature of life, the fact that life requires integration with other life to exist, provides a practical side to the idea of the ghedeist. We need to know what direction to go in at times, and the ancient "good" in the ghedeist provides a measure of guidance. Remember, "ghehd" has meanings connected to uniting, joining, gathering, and good. This is precisely what symbiotic science has discovered as a foundation for life in the twenty-first century.

In symbiotic science, we have seen that cooperation extracts from the Earth as much as is possible to expand and extend life (to grow) and to defeat the forces that will pull things apart and cause sickness and death. Life, as they say, is a strategy that opposes entropy and decay. There is no such thing as waste in the symbi,oment of life. Everything that is alive, or was once alive, is consumed and "recycled" as life. Fungi seek out and share scarce elements and minerals that plant roots cannot find; bacteria accept and help process our food and, in return, get nutriments, regulate blood pressure, and protect us from the bacteria that would otherwise destroy us.

The great ethical drama of terraphthora and terranascia is the story of life. For life to endure, terranascia must prevail in the long run, despite all the forces of random disaster and calculated predation aligned against it. The vicissitudes of life can include the asteroid hit tomorrow, a chance encounter with a human-eating tiger, or coming into contact with flesh-eating bacteria. These events and creatures are undoubtedly bad for those whose life is extinguished by them. They are rare events, however, compared to the overwhelming number and scale of terranascient events going on in the normal course of life. After the disaster, when life resets, it continues to build complexity and united diversity. Terranascient forces are good, terraphthoran forces are bad. Terranascient forces work together to protect life in symbiotic relationships. Terraphthoran forces not only attempt to destroy other life forms, they will also turn on each other.

We already have a context where this type of persistence is to be found. It is the way the human mind works. The Australian philosopher John Anderson explains what I consider to be the parallels between the working of life and the functioning of the human mind:

> But the same may take place within one person's mind, when a conflict is resolved and a new type of activity emerges by the aid of certain abiding motives or sentiments. This is the process of *sublimation*, where one motive finds for another a means of expression, provides it with a language, puts its own *ideas* before it as objectives. This is also the process of education. It may be argued, then, that all good motives have this power of transference or conversion, whereby from hitherto dissociated material a new motive is formed which can cooperate with the good motive. Goodness is associative, evil is dissociative; goods have a common language, evils have not.[45]

I disagree with Anderson in that, in the spirit of Freud, evils do have a common language, but it is the language of isolation and death. Thanatos, necrophilia, and terraphthora are evil triplets. However, the death instinct and necrophilia are dead ends in both the symbioment and the mind. Their ascendancy is always short term.

In addition to personal contexts, both virtue and evil can be found manifest in social movements where individuals cooperate with each other to achieve common ends. Virtue achieves progress without the repression of others, while evil makes temporary "progress" only at the expense of others and, in the long term, itself. Evil perpetually fails the test of universalizability, since there are always contradictions and conflict evident in its application. It is difficult to conceive of society (any society) continuing to exist if evils dominated good in the long term. As Kamenka argues, goods "cooperate with each other and display internal progress and development in a way that evils cannot cooperate and progress."[46] In this doctrine there can be hope that dissociative and antilife forces in politics will end up fighting each other and leave associative, soliphilic forces to get on with the business of growing life. It is the basis for optimism.

Thus, in the global order there is a parallel with the associative and dissociative qualities present in psychosocial good and evil. Ethics, in the symbiomental context, is concerned with the linking of associative psychosocial attitudes and institutional design with the patterns of symbiotically derived order to be found within life systems. These organic connections may occur in personal reflection, social movements, and ultimately institutions that are self-determined, self-organized, and self-perpetuating forms of symbiotic organic order. Allow these elements free rein, and you get the Symbiocene.

The idea that organic unity and order pervade all life in the biosphere brings with it the recognition that the ethical or the good is built into the objective order of things at a biome scale. To identify the potential for good (associative, organic, symbiotic order) from evil (dissociative, exploitative, parasitic disorder) in the global context is the most important task ethics can undertake.

Ethics is the social equivalent of the foundational symbiotic relationships generated in other parts of the symbioment; it is a social strategy that opposes social entropy and disorder. Such a strategy becomes increasingly important as humans move from a tribalized, patch-disturbing, and local species to an interconnected global species, and then, I will argue, into Symbiocene tribes within the Symbiocene (see chapter 6).

While our evolutionary heritage may have required the dissociative in intra- and interspecies relationships at times, it must also have had a central place for the cooperative and the associative (sex, reproduction, nurturing, education). The stakes are now too high to allow the parasitic and selfish view of evolution to assert "hard-wired" individualism over symbiotically acquired cooperation and mutual aid. The generation of positive, associative, ghedeistual-organicist, and fluministic ethics from within a globally affiliated human culture needs to be put in clear contrast to its selfish, terraphthoric alternative.[47]

As one of the best-selling box office films of all time, *Avatar* looks increasingly like a documentary on the human plight on Earth. Ecological and biological science continue to build the case that the condition we call life has, at its foundation, grand-scale symbiosis within organisms and between species. Without symbiosis, complex life as we know it could not exist. Interconnectedness and sharing are the norm, isolation and disconnectedness are harbingers of death.

We have seen that the characteristics of the fictional moon Pandora actually have a scientific foundation, made more explicit in scientific discoveries in plant-plant communication, plant-fungal networks, and human gut biome research, since the film was made. The spiritual and lovelike qualities of the Na'vi nature were based entirely on a naturalistic ecosystem that also reflected that sense of interconnected life and love. Their social life was based on sumbiophilia and soliphilia—the cultural and "political" expression of symbiosis. James Cameron had created for them their very own proto-Symbiocene.

Humans, biologically attuned to such a uniting life force, felt the emotions of connecting with it and, when the film was over, detachment from its beautiful attraction. Such emotional feelings are sharpened when we have the grim realization that Pandora is a fiction and that, so far, we have found no evidence whatsoever that life exists anywhere else other than right here on Earth.

There is a similar process happening with the Anthropocene. As its all-enveloping power extinguishes and smothers life on Earth, the life form we call "human" is experiencing a dissociative invasion. This time it is not human versus Na'vi, but human versus humans (and the rest of life), all in the quest for "unobtainium." We are a species at war with ourselves and all living forms with which we share the Earth. There is a shattering of the common life spirit, and I argue that we all feel this shattering in various ways. The negative Earth emotions felt in people are all symptoms of life being torn apart, global *koyaanisqqatsi*, or life dis-integrating.

Along with the material collapse of life and its vitality, there is also the collapse of what we once would have called a spiritual dimension. As each shared intersection in the ghedeist of life is extinguished, as each holobiont biocomunen is wiped out, we experience a diminishment in our personal and collective identity. As the sixth great extinction proceeds, we sense that we are getting closer to the seventh great extinction . . . ourselves. I feel this loss in my integrity when the last West African Black Rhinoceros is shot for its horn, or when the body of a little boy, Alan Kurdi, a refugee, washes up alone on the shores of the Mediterranean Sea. The ghedeist connects us all.

Organized religion has never had a mortgage on spirituality, but it has been a powerful broker of its meaning in the past. Belatedly, as many forms of nonhuman life disappear from the face of the Earth, the leaders of many faiths have called for greater efforts to conserve and protect what we have left of "creation." Pope Francis is a great leader of the Catholic Church in this respect. In my view, however, spirituality, as the human search for meaning, has already left the big tree, the big church, and the "big man" view of life. It has returned to where it all started: collaboration and association among the smallest possible units of life. That life can grow and evolve from such humble beginnings, to now have its human form discover how it all happened, is itself spiritual. It gives generously of meaning as to how humans ought to live.

The symbiotic turn in our thinking takes us out of formal religion, out of reductionism, out of patriarchy, out of economic liberalism, out of auto-poiesis, and out of the idea of the autonomous individual.[48] Finally, we are liberated from the dominant ideas that have cajoled us into thinking that we are separate, or must separate, from the rest of individual life forms. We can no longer excuse or justify that separation. Life is a cooperative enterprise held together by associations of diverse life forms that have coevolved over millions of years. Humans are simply a recent example of large-scale symbiosis in action, as we walk this Earth courtesy of our gut bacteria.

To have spirituality without religion might come as a relief to many. I feel it is long overdue, as doctrinal spirit in the realm of the otherworldly has now been replaced by the ghedeist within this world, in every nook and cranny. The ghedeist is in every living thing and, if you are open to it, you can feel it everywhere and within yourself. It is about time we loved what and who we are.

Rock art in Arnhem Land, Australia.

Photograph by the author with permission from traditional owners.

Chapter 6

Generation Symbiocene

Creating the New World

The generation of humans known as baby boomers is, by and large, a lost cause for entry into the Symbiocene.[1] With entrenched beliefs, and cognitive dissonance about the non-sumbiosic path that has led to our present predicament, and having accumulated wealth, power, and privilege never before seen on a mass scale in human history, boomers are not going to give up their beliefs or their wealth anytime soon. They would rather adapt to changing circumstances than mitigate, or fundamentally change, any of the foundations of the Anthropocene.

Boomers "have it all," including an inflated notion of the self, but others, including their own children, are now paying the real cost of such stolen wealth, with their deeply entrenched personal, social, and environmental "problems." Tim Flannery called this process "future eating."[2] A tougher description could be conceptual and ecological cannibalism. Given that I am a baby boomer, I hope my pronouncements are wrong—totally wrong.[3]

The post-boomer children, which now include the generations X, Y, and Z (all three henceforth shall be collectively known as the "Gens"), are all depicted to some extent in popular literature as suffering various forms of sociological and psychological problems, including being uncommitted, unfocused, directionless, and dependent and lacking a coherent identity.

Locked into gaming, entertainment, and social media, they are described as not connected to external reality, spending most of their time in the virtual world and indoors. Gen Z, born between 1995 and 2012, might be on a different track; they are characterized as having a global perspective and being more critical of social media than their predecessors. I am sure these marketing-generated generalizations are unfair to many Gens and I will refer to their positive attributes later in this chapter. However, let us go with the marketing generalizations for a while and see where they take us.

In the previous chapters, I have argued that in order to enhance positive Earth emotions, getting out of the toxic Anthropocene is vital, and that in addition to the scientific and ethical foundations of the Symbiocene, a secular "spiritual" foundation in the form of the ghedeist was also needed. In this chapter, I will provide additional theoretical and practical reasons for a new, positive direction for all generations of humans on the planet. In order to get out of the Anthropocene, humans will need generational change. I will make the case that Generation Symbiocene (Gen S) will arise from within the ranks of all of the post-boomer generations (Gens) as the need for a new human identity, built on common Symbiocene principles, politics, and ethics, becomes clear.

The Anthropocene Identity Crisis

The Anthropocene is becoming increasingly "rational" at achieving a totally irrational end. Under the logic of gigantism and homogeneity, we are destroying the very economy we are trying to build. As value in the world is converted to increasing shareholder profit, everything else becomes impoverished and all forms of heritage are lost.[4] Economies need consumers, but consumers need jobs to purchase consumer items. As we make production more efficient, cheaper, and automated with artificial intelligence (AI), big machinery, and robotics, there are fewer jobs in all forms of human enterprise. Farms become automated and farmers disappear. Lawyers are replaced by robots programmed full of case law. Accountants' work becomes automated or is "blockchain" bypassed, so we need fewer accountants. Self-driving cars, buses, and trucks eliminate drivers. AI, created by programmers, eliminates many programming jobs in the world of computing.

On top of losing employee-consumers to this irrational system, we are losing the Earth by contamination of the air, water, and soil. And we appear to be allowing politicians and private corporations to do this to us and our world. I cannot imagine a scenario that is more irrational. Taken to its "illogical" conclusion, we have a weird world where wealth and power are increasingly concentrated at the top of a social hierarchy, while all the negatives of that wealth accumulation are felt by the massive majority of people, all over the world, at the bottom. It must also be said that nonhuman beings are also feeling negatives such as endangerment and in some instances extinction in the wild. There is a clear relationship between the extinction of jobs and the extinction of species since, as more people become unemployed and displaced by AI and automation, they become less concerned about novel job-creating exploitation of what remains of the undeveloped world. The rise of the fly-in fly-out worker exemplifies this trend as, for example, much mining employment in Australia now takes place in remote parts of the continent and the workers fly back to "first-world" suburban locations effectively immune to the negative impacts on that remote site of exploitation.

Yet I have to look on the bright side of this irrationality, or what I call artificial stupidity (AS). There is tension building here that has all the characteristics of a really good psychic and emotional crisis, precipitating a conceptual and emotional revolution, which may avert the biophysical meltdown of the planet. We are already in a stage of psychic upheaval that will lead to this sumbiorevolution.

As the corporate giants become even more bloated, they become vulnerable to total failure. A simple rule of complexity theory is that the more diverse a system, the more likely it is to be stable. Failure in one part does not entail failure in all other parts. There is built-in resistance to waves of perturbation (change), a condition technically known as "resilience." Conversely, the more uniform and rigid the system, the likelier it is that any failure will deliver total failure. As corporations and governments become increasingly dependent on the World Wide Web for logistics and accounting, they open themselves up to total system hacking and failure. Vulnerability to such failure is now one of the hallmarks of the modern quasi-monopolistic corporation.

Finally, global efforts to combat climate warming and other environmental insults suffered by the Earth all came to a grinding halt in late 2017. The United States has threatened to pull out of the Paris Agreement (2015) on greenhouse gas emissions and is now on a path of hugely increased carbon emissions. Despite almost thirty years of intergovernmental effort and negotiation on lowering global emissions and reducing the concentration of greenhouse gases in the atmosphere, they are still rising. The goal of keeping the world under two degrees of temperature rise by 2100 appears now to be impossible.

Assessing climate change threats to identity is crucial, since current science-based prediction of, for example, ice melt in Antarctica, can deliver up to three yards/meters of sea level rise by 2100.[5] Blaming the victims of sea-level rise in places like Bangladesh is not a solution. Attempting to keep them out of other countries—by building walls, for example—will not work either, as the numbers of people who are made refugees by all forms of climate chaos, should they emerge, could rise to 2 billion by 2100.[6]

After the global financial crisis, which began in 2007–8, the world's financial system has been increasingly understood as extremely fragile because it is largely built on debt. A lot of that debt became publicly funded bailout money from governments. The global Occupy movement in 2011 questioned this irrational system for over a year but lost momentum as people became tired of harassment by law enforcement, distracted by Netflix or somebody's cat on YouTube. The problem, however, has not gone away, and the global financial system remains at a tipping point. With the ongoing fiscal crisis, plus climate chaos, the economic and biophysical crises faced by all of the Gens cannot be ignored any longer.

I am of the view that all the Gens will quickly see that the damage being done by disaster economics and human-enhanced unnatural disasters is increasing each year. Ironically, the United States is one of the countries being hardest hit by the frequency and scale of these disasters. The Wizard of Oz could not hide behind his curtain of lies for very long, and neither will the current batch of political leaders. The truth about fiscal irrationality and the insanity of deliberate climate warming will shock all the generations, and they will enter all forms of negative psychoterratic states. It will shake the foundations of current human identity to its core.

A simultaneous ecological, economic, psychological, ghedeistual, and cultural collapse has never before been felt by humans on a global scale. There are massive contradictions that have global origins, tearing our societies apart, but they are being felt locally and regionally by people in multiple forms of trouble. It is no wonder that populism, with its slogans and the offer of simple solutions to incredibly complex problems, is the first port of call for those who are worried about the future but have no idea about what is building such pressure in the present. Many are ready to blame the victims of such forces rather than address the complexity. On top of all that, there is the world of false facts (lies), false data (lies), and fake news (lies) that people have to contend with. Even those who are supposedly leading intellectuals and policy experts apparently find it very hard to discern fact from fiction.

We need to be open about the fact that negative human-induced impacts on Earth are on the rise in the age of solastalgia. Every significant indicator of our relationship to the biosphere is heading in the wrong direction, and world scientists have collectively given us clear warning.[7] Along with these negative biophysical indicators come a host of negative, nonmedical, psychoterratic states, ranging from mild forms of anxiety about the present to solastalgia about the lived experience of negative change and deep dread about an ecoapocalyptic future. Diagnoses of depression and of an expanding number of serious disabilities have made mental health one of the largest areas of growth and expenditure in the total health budgets of nations.[8]

There are some, including many leading public figures and intellectuals, who think that the situation is already so bad that humans should seriously consider leaving the Earth to colonize new planets. They suggest, for example, that the mining of virgin resources on nearby asteroids can sustain humans back on Earth. Jeff Bezos, the founder of Amazon, has stated that "we have to go to space to save Earth. . . . We kind of have to hurry."[9] Other well-known men, such as Brian Cox, Elon Musk, Richard Branson, and the late Stephen Hawking, see the tierracide coming and speculate that space-colonizing refugees will be the only survivors of mass economic and agricultural failure on Earth. Hawking concluded: "Spreading out may be the only thing that saves us from ourselves. I am convinced that humans need to leave Earth."[10]

According to this current crop of space invaders, while some of us are still rich, we must plan and execute our escape from a failing Earth and its failing humans. What an irony it is that, as Jeff Bezos becomes the richest man on Earth and Amazon expands worldwide, the Amazon Basin is contracting at record rates. These public figures in science and business are in a position to be intellectual leaders, yet their strong advocacy of the extraterrestrial, at a time when "terra," the Earth, is in big trouble, marks them all as in need of strong challenge on ethical grounds.[11] The current "extraterrestrial" patriarch pack of the Anthropocene are failing us. Projections for the future of the Earth, from the impacts of climate change alone, suggest to me there is no justification for directing huge amounts of wealth, public or private, into anything but the total mitigation of the threat of climate chaos in its various forms of Armageddon (flood, heat, disease, and tempest).

The US writer Derrick Jensen has put the case that failure to appreciate the amazing life on this Earth marks the patriarchal space invaders, some of whom are scientists, as suffering from a form of identity pathology:

> With all the world at stake I need to speak plainly. The problem is that within this patriarchy, identity itself is based on violation. Violation becomes not

merely an action but an identity: who you are, and how you and society define who you are. Within this patriarchy men's masculinity defines itself by identifying others—any and all others—as inferior (which is why those stupid fucking scientists can ask "Are we all alone?" as they destroy the extraordinary life on this planet), and as being therefore violable, and then violating them.[12]

Following Jensen, central issues for boomers and subsequent generations are the questions of personal and political identity. Personal identity based on "violation," abandonment, exploitation, and power has become a defining theme in the early twenty-first century.[13]

The key markers of who you are and what you believe in are under constant construction, with feedback loops providing the confirmation needed for identity security. The standard identity makers, such as genes, parental influences, school, and role models, continue to be important in the development of the personality and identity. However, they are diminishing in relevance under the power of new media and their connection to information, propaganda, consumption, and advertising.

While contemporary social media and other technologies can provide outlets for the personality, they require a form of obedience to formats and formulas that are predetermined by those who own and control the media and its commercial products. The nuances of personality are lost in an e-world, where to "like" and use various one-click "emoticons" are seen as adequate responses to hugely complex issues such as global warming. Popular social media platform Twitter limited the length of individual comments to 140 characters per tweet (in late 2017, this changed to 280), and within such a constrained format no nuance of dialogue and argumentation is possible. The anonymity allowed in most formats also works to hide real personal identity and to give free rein to insulting and discriminatory utterances. Fake news becomes hard to identify in a world where everything is brief and without verifiable sources. Facebook might offer the possibility of many "friends," yet the owner of a personal page might still be intensely lonely in the real world.

Social media identity, based on the principle that you are what you consume and what you "like," is now carefully crafted around algorithms that anticipate your every next move, and put before you choices that reflect your past behavior. Your "likes" become circular in that the choice of who you are is determined by what you have been. Little room for identity innovation here, or for critical evaluation of identity culture, economics, and politics.

The tendency to join like-minded e-groups also ensures that the "echo chamber" effect, or the search for self-confirmation in friendly confluences, reinforces one's own position on matters of fact and value.[14] The current system "grooms" you on a constant basis to ultimately do what it wants you to do: consume more of what you "like" and become more compliant and more predictable. It also gives constant positive reinforcement as to the "correctness" of your views, as dissenting opinion is filtered out. The final insult is that the owners of major social media corporations can sell your personal data to third parties for political and commercial interest. Privacy can now be sold, as was pointedly illustrated by the revelations about the data-selling company Cambridge Analytica in 2018.

Generation Identity and the Rise of Nationalism

Partially as a result of the characteristics of social media, identity politics linked to renewed nationalism and the core issue of "violation" have emerged in the last few years as responses to waves of refugees seeking asylum in various parts of the world. Muslim refugees, escaping war in the Middle East, have become the focus of anti-Muslin sentiment in the United States, Australia, and parts of Europe. The perceived threat to cultural and political identity in the minds of many in the receiving countries has become the reason for a revival of nationalism and identity marking based on what might be considered the traditional values of the people of a particular nation. The role of new media in helping form and promote these neonationalist and cultural identity politics has been crucial in their expansion during 2017. For example, a group known as Generation Identity (Gen I), based initially in Europe, has taken over ground that covers the following:

> Generation Identity (in some countries running as "Identitarian Movement") is a Europe-wide patriotic youth movement that promotes the values of homeland, freedom and tradition through peaceful activism, political education, and community & cultural activities. We want to create an awareness for a patriotic value base in the metapolitical realm.[15]

While claiming on their web site *not* to be neo-Nazi or fascist, this youth group has appealed to many thousands of young people in Europe who oppose, among other things, refugee intake from countries that are mainly Muslim.[16] They present themselves as patriots for their own country, view immigration as a net impost on society, and are strong advocates for remigration of refugees to their home country, as they vigorously oppose what they call the "the Islamization of Europe." There is an overall emphasis on controlling national borders and a reemergence of the concept of "sovereignty."

In rejecting a cosmopolitanism identity and globalization as its path, Gen I has hit on something important, but has mistaken national identity and patriotism in the nation-state as the focus for meaning and solidarity with others and has completely ignored the ecological foundations of place.

The nation-state can be a powerful force in motivating strong allegiance to place, but as has been shown in the twentieth century, once such allegiances become entwined in racism and ethnic issues, they can very quickly degenerate into mass violence, genocide, and the obscenity of "ethnic cleansing." Extreme forms of nationalism, such as those demonstrated under the names of National Socialism (the German Nazi movement and the Third Reich) and Fascism (Italy under Mussolini), are good case studies in what not to do in the name of identity. Two world wars and the attempted genocide of minority groups are not paths to follow in the quest for identity in the twenty-first century. Besides, terraphthoran National Socialism and Fascism lost both wars. I suggest that extreme nationalism and war are not the sociopolitical paths that the majority of the post-boomer generations wish to take in the early twenty-first century.

War is undesirable for another reason. At times of interstate war and intrastate conflict, the protection of land and biodiversity is not normally on the agenda of the combatants. There is so much "collateral" damage on those who are in harm's way. Ecocide, generally, is an inevitable consequence of war and genocide. The understanding and protection of the ecology of the local or regional natural environment does not even appear on Gen I's major list of demands.

Xenophobia, racism, injustice, gross inequality, and anti-environmentalism are a toxic blend. In a world that is rapidly changing it is important to examine whether there is a compelling alternative to identity based on nationalistic atavism and ecocidal notions. Given the actions and outcomes of European imperialism and colonialism over five hundred years, the globalization of industrial capitalism over the last two hundred, and mass migration due to world war and other forms of conflict over the last hundred, it seems absurd to lapse into some notion of ethnic or racial purity or superiority to generate contemporary identity. There is only one human species, and separate "pure" races are a myth easily unmasked by genetic testing of human DNA. The majority of countries, all over the world, have populations that express both the diversifying forces of colonialism and migration. There is no way that these cultural melting pots can be undone. Adding to the historical melting pot are the likely future impacts, as noted above, of forced migration due to climate chaos.

A similar argument can be made for the nonhuman symbioment, or what now remains of it. I have made the point about the possible end of endemism in a previous chapter, and although we are not at that point right now, we are heading rapidly toward it. The global mixing of micro-organisms, fungi, plants, and animals has taken place rapidly in the last two hundred years, and there is no undoing that melting pot either. Climate warming also makes that problem worse, as plants and animals are forced to migrate, either toward the poles or into higher altitudes, to stay within their preferred ecological-thermal niche. Humans are already involved in determining their future, and will be even more so as relocation options, such as moving Polar Bears and whole Arctic food chains to Antarctica, are considered.[17]

In thinking about identity in these contexts, we now have to accept the reality of emergent hybrid cultures and emergent hybrid ecosystems. Our speculation about a terranascient future for the rest of this century must accept the dynamic reality of place. This may mean accepting the impoverishment of ecocultural diversity from a more diverse benchmark of only a few decades ago.

Using as foundations older traditions in culture, science-based definitions of biophysical regions (soil types, vegetation etc.), plus new biomic science, humans will need to redefine their emergent identity status in opposition to the now-failing globalization and nationalization in the Anthropocene. That task, especially in the light of the impacts of climate change on local and regional ecologies, will not be easy, but it is still the case that geography plays an important role in what humans can do in specific locations on Earth.

The application of Symbiocene principles (chapter 4) will see the gradual reemergence of cultural and ecological endemism in different places on Earth over time, and I will say more on this theme below. Before doing that, however, I think it worthwhile to see if there is anything worth salvaging from past conceptions of country, nationalism, and regionalism with respect to identity, that could help the Gens move into the Symbiocene.

Identity Fragments

As I argued in the previous chapter, the traditional Aboriginal society of Australia satisfied some of the Symbiocene principles. Their close allegiance to what we now understand as "country" was, for them, based on intimate knowledge of place, love of place, education about place and its history, and a desire to defend their place from invaders. Australian Aboriginal people still have powerful affiliation and identity with their traditional places of birth. Many Australian Aboriginal people prefer not to be identified as Indigenous, indigenous, aboriginal, or Aboriginal, but wish to be known by the traditional tribal name of their country. Such is the depth of their regional and cultural affiliation to place. Australian Aborigines are not unusual in having a love of country and a suspicion of invading strangers. This is especially so when those strangers turn out to be representatives of aggressive colonial powers.

Despite that reservation, Australian Aborigines have a long record of willingly incorporating new people into their territory and culture. In post-1788 Australia, white people, lost castaways, and escaped convicts in the early settlement were often taken in and cared for by Aboriginal people. They were given tribal names and lived within the rules and conventions of their clan.[18] When strangers are willing to learn and respect the ways of the traditional people, they are accepted. There is also a rich exchange of knowledge, and an emergent hybrid culture is the result.

Aboriginal Australians are also generous in their incorporation of introduced, feral animals into their belief systems. The Asian Water Buffalo, for example, has been accepted by the people of Arnhem Land into their Dreaming because this animal, as a water-based shaper of the landscape, has affinity with the Rainbow Serpent of the Dreaming, responsible for carving out serpentine water courses and billabongs.[19]

The spirit of generosity and caring for life suggests to me that human identity can be tied to particular regions, but without the extreme xenophobia that has been seen in the past decade or more in Europe and the United States.

There are many other examples, worldwide, of indigenous groups based on soliphilia, representing some of the Symbiocene principles discussed in chapter 4, that have defended long-term and viable connections to their place/country. The Standing Rock Sioux Tribe, with the support of many environmental and civic interest groups, opposed the Dakota Access Pipeline in North Dakota in 2016, on the basis that it violated a sacred burial ground and had the potential to leak and contaminate drinking water supplies. They maintained that their special interest in traditionally owned land trumped the need and the authority to build the pipeline. The Standing Rock Sioux, and their support group, successfully delayed the construction of the pipeline until March 2017, when the US president authorized its construction and the dismantling of the protest camp. As was the case with the Occupy movement in 2011, the opposition to the pipeline was emotionally charged but quickly thwarted by the authoritarian forces working for the nation-state, globalization, and gigantism. Resistance was crushed as leaders got arrested and some were jailed. Maintaining traditional identity in the face of all the forces of a nation-state that opposes you is a very difficult position to take.

The resistance to regionalism or micronationalism by the nation-state has also been demonstrated in recent history. We have seen a long-term secession movement in the Basque region of Spain and, in 2017, within Barcelona in the region of Catalonia. Both secession movements have been strongly resisted by the Spanish government. While recognizing racism and anti-immigration as primary factors, asserting one's own unique identity in the face of strongly homogenizing forces (e.g., the European Union and the UK government) was also an issue in the Brexit vote in Great Britain in 2017. Prior to that, there has been a revival of Scottish nationalism in the UK, and Wales has always been a place that celebrated its regional difference from the rest of Great Britain. There are even occasional secession movements in Western Australia, since geographic isolation from the other states of Australia, its own distinct way of doing things, an economy that works independently of the rest of the continent, plus a highly endemic biogeography, combine to assert independence from the rest of Australia. Similar "secessionist" tendencies have been seen in the United States, with "Cascadia" based on the bioregional affinities among West Coast states and a revived Calexit movement in California, after the election of President Trump in 2016.[20]

One of the reasons for the appeal of secessionist movements is that people with long ties to distinct bioregions have regional dialects, culture, and agriculture, and strongly identify with them. Many love their landscapes and have endemophilia for their distinctive features. Even in contemporary Europe these particular and unique regions remain, and the people within them continue to contest their right to self-determination against the claims of nation-states and broader superregional and global forces. It must be admitted, however, that the idea of a secessionist identity is not on any mainstream agenda right now. The nation-state seems firmly entrenched, especially when it is supported by the interests that run multinational capitalism.

The bioregional movement, popular in academic and environmental circles in the 1970s and '80s, actively opposed gigantism in its advocacy of the right scale for human life.[21] Kirkpatrick Sale saw all aspects of human potential, including identity, as maximized within the natural characteristic of a region. He argued:

> At the right scale human potential is unleashed, human comprehension magnified, human accomplishment multiplied. I would argue that the optimum scale is the bioregional, not so small as to be powerless and impoverished, not so large as to be ponderous and impervious, a scale at which at last, human potential can match ecological reality.[22]

The bioregional movement expressed the hope that humans can come to know and love their bioregion, live within its particular features and properties, and govern themselves. This is an idea that celebrates ecological terroir and its cultural equivalents, and opposes the gigantism of globalization. Sale, other bioregionalists, and social ecologists such as Murray Bookchin have argued that as ecology, culture, and politics are invaded and metastasized by global-scale development forces, the Earth becomes impoverished and open to catastrophic change. The transition occurs because those who know their symbioment or bioregion and know how to nurture its sumbiosity, are pushed aside from governance by more powerful hierarchical and homogenizing forces, which are not only often absent from the bioregion, but are ignorant of its special features and characteristics. It gets worse than that when corporate powers are ignorant of global-scale life-support systems, such as water resources or the climate, which in turn support bioregions. Then, on local and global fronts, heterogeneity is lost while the homogeneous forces of a universal model of development, capitalism, become seemingly unstoppable.

Unfortunately for the bioregional movement, despite the efforts of people like Sale and Bookchin, the push toward gigantism, not bioregionalism, has exponentially increased over the last fifty years. Even Perth, Western Australia, the most isolated capital city in the world, was not immune to the forces of gigantism and globalization. When working in that state from 2009 to 2014, I experienced both nostalgia and solastalgia for the place of my birth. I mentioned in the sumbiography that my identity was tied to that landscape, and watching it being bulldozed for yet more suburbia felt like an attack on my emotional core, my identity. Moreover, while resident, I also experienced an episode of climate warming–induced mass tree die-off in the Jarrah forest where I lived. I wrote in my blog in April 2011:

> I live in the hills at Jarrahdale and well into April I am watching whole sections of the forest slowly die. It is not just Jarrah and Marri that are turning yellow and dying, the whole ecosystem is in deep distress, so much so that, tragically, it looks like a Northern Hemisphere autumn is taking place right here, right now.
>
> If we could watch what is going on in fast forward we would see vast tracts of bushland dying of thirst in the grip of this permanent drought. If you take the time to look, you will notice that the native ecosystems on the coastal plain are also in the deep distress of various forms of "dieback." What is happening is a "tipping point" in the process of "tipping" . . . a rare but hugely important event.
>
> The Perth Region, including the Hills, has had a 20% decline in rainfall over the last 30 years and a much larger, 60% decline in runoff into the streams and our dams. Gooralong Brook, once a year-round running stream, has disappeared and most of its deeper pools are now bone dry. The climate of the SW of WA has already changed for the worse and if it gets even warmer and dryer, the Perth region will be in perpetual ecosystem distress.
>
> This distress is not only about trees, frogs, jilgies and thirsty kangaroos; it is also a crisis of the human spirit and the mind. Our identity as people of the Perth region is at stake. All that is endemic to this special part of the world is at risk of slow death by desiccation. Our iconic trees such as Banksia and Jarrah are already dying and the wildflowers, the exquisite ground orchids and kangaroo paws, will not reappear in a dry, colourless Spring.

The Perth case study not only represents the rapid diminution and loss of identity for baby boomers such as myself, but it also marks a graphic case of environmental generational amnesia or the compounding ignorance (eco-agnosy) for all generations after the boomers.[23] Each generation knows less about their bioregion as an outcome of cumulative environmental desolation. The so-called environmental crisis experienced as a loss of diversity is also ultimately a human identity crisis. These twin crises forced me to think that while bioregionalism is a wonderful idea, this was not the kind of idea that would generate an identity revolution. The deep-seated features of bioregions, as envisaged by writers such as Sale and Bookchin, were being erased from the maps of urbanizing and industrializing locations all over the Earth. Gradual removal of the endemic was taking place, while at the same time new forms of invasive heterogeneity from all over the world were being introduced. The net result for people and place was a gradual superficial blending of culture, ecology, and a built environment that is almost indistinguishable from anywhere else on Earth and certainly anywhere else within the nation-state.

Similar ideas about scale and homogeneity and their relationship to identity were presented in 1992 by Benjamin Barber in "Jihad vs. McWorld," where "McWorld" represented the negative forces of global-scale homogeneity of corporations and the national state, and "Jihad" represented the autonomous negative forces of tribalism, including religion, ethnicity, and some revolutionary nationalist politics. Both positions, Barber argued, were deeply flawed, with McWorld enforcing a diversity-killing uniformity on the world, and Jihad killing (often literally) everything that is not itself. The net result is the same: enforced uniformity caused, ironically, by each other's presence. Barber argued:

> The tendencies of what I am here calling the forces of Jihad and the forces of McWorld operate with equal strength in opposite directions, the one driven by parochial hatreds, the other by universalizing markets, the one re-creating ancient subnational and ethnic borders from within, the other making national borders porous from without. They have one thing in common: neither offers much hope to citizens looking for practical ways to govern themselves democratically.[24]

Barber's book was prescient in that actual Jihad within Islam has become a major threat to the security of nation-states worldwide in the first two decades of the twenty-first century. The emergence of Generation Identity and explicit pronationalist movements in Europe (Poland and Austria in particular) has also conformed to his hypothesis. At the same time, the forces of capitalist global gigantism have become even more threatening, as population, industrialization, and consumption have increased exponentially. Barber could see the injustices to people and the environment building over time. He suggested:

> The impact of globalization on ecology is a cliché even to world leaders who ignore it. We know well enough that the German forests can be destroyed by Swiss and Italians driving gas-guzzlers fueled by leaded gas. We also know that the planet can be asphyxiated by greenhouse gases because Brazilian

farmers want to be part of the twentieth century and are burning down tropical rain forests to clear a little land to plough, and because Indonesians make a living out of converting their lush jungle into toothpicks for fastidious Japanese diners, upsetting the delicate oxygen balance and in effect puncturing our global lungs. Yet this ecological consciousness has meant not only greater awareness but also greater inequality, as modernized nations try to slam the door behind them, saying to developing nations, "The world cannot afford your modernization; ours has wrung it dry!"[25]

The significance of climate warming has hugely escalated since Barber's hypothesis in 1992, and we are now facing the multiple crises of gross inequality, psychic upheaval, gigantism under globalization, and climate chaos all poised at tipping points of geopolitical stability. While they are stories of oppression and ultimately limited success, indigenous movements, secessionist movements, bioregionalism, and jihad have within them elements that can be used to construct a regenerated version of human identity, one that repossesses place from all forms of colonialism and liberates it from universal forms of oppression and tierracide.[26]

Rebuilding Identity and Constructing the Symbiocene

From this point onward, the human world must be restructured to simultaneously address identity, inequality, and ecological destruction. Capitalism attacks identity on all fronts by homogenizing all before it. As Naomi Klein has effectively argued, the climate contradiction has reached the point where "this changes everything," including the so-far achieved homogenization of human identity under McWorld.[27] In resisting that tendency, there is potential for unity rather than jihad among all those who value diversity. Such resistance goes beyond all previous class-based acceptance or rejection of capitalism. It also goes beyond all previous forms of religion and philosophy. This is unprecedented. Some new symbiotic form of reasserted place identity and identity politics has to become more important than everything else.

It will be the recognition of the centrality and vital importance of the sumbiosic at the local and regional level that will enable the innovation required to usher in and perpetuate the Symbiocene. Preserving diversity and satisfying the Symbiocene principles require an economics not yet understood by capitalist and socialist gigantism, even in their most ecological of forms. As Herman Daly has argued:

> Different nations follow different policies: some succeed, others fail, and presumably all could learn from each other's experience. In an integrated globe, there is one grand policy experiment, so failure has huge consequences. In evolutionary terms, there can be no group selection with only one group—success and failure are blended within the whole rather than separated by groups. Furthermore, cultures really are different, and what works in one country might not be right for another, much less for all. People exist as members of national, linguistic, cultural, and religious communities, not as global secular cosmopolitan individuals all speaking Esperanto at Davos. In other words, cosmopolitanism seems to be assumed in global ecosocialism as well as in global capitalism, while the legitimate claims of communitarianism are relatively neglected in both cases. Likewise, the larger scale of organization seems to be privileged over the local. By contrast, Reinhold Niebuhr thought that "the larger the group, the more certainly will it express itself selfishly in the total human community." This would suggest that even the nation-state is usually too large.[28]

With a chronic change agent such as climate warming, there is still time for people within bioregions to assess and document the impact of such change on their landscape and biota. Climate impacts on agriculture are also evident over time. With these chronic changes, plus the cumulative impact of the increasing frequency of flood and storm damage, citizens become increasingly aware of the sheer importance of local and regional landscapes for their own survival and safety. Even in cities and towns, especially those that have frontage to open ocean or are close to rivers, regular flooding and damage add to the awareness of unwelcome biophysical change.

As negative change occurs to loved and vitally needed places, it reasserts the primary importance of the biophysical in the lives of humans the world over. That awareness will be sharpened by the possibilities of mass migration into "safe" areas or mass migration out of "unsafe" places. Studies have only recently projected the impacts of these mass movements of people in locations such as the United States.[29] As the risks and actual disasters ratchet up, people will have not only increased negative psychoterratic feelings such as ecoanxiety and solastalgia, but also emergency engagement of positive Earth emotions such as topophilia, endemophilia, ecophilia, and eutierria. It is also likely that people will politically unite in soliphilia as a local response to the threats. As a resurgence of local and regional identity takes place, I think it would be a mistake to see such a resurgence as simply an expression of crude xenophobia for others or emergent fascism or both. On the contrary, resurgent soliphilia-based political action becomes a sure sign that sumbiocracy for the Symbiocene has commenced.

Another factor in a possible revival of localism and regionalism is that at least 3 billion people live and work locally, away from city conurbations, and are not highly mobile global citizens. While they might use basic cell or mobile phones, they do not have an e-identity crisis, as the globally connected smart phone and personal computer are still economic and technical impossibilities in most parts of, for example, Africa. We need to remember that about half of the world's population still lives in a small town or rural village and is mainly sustained by its hinterland. These people are already intensely local in their survival orientation and will be highly motivated to protect their patch should the need arise.

Ursula Heise has made a strong case for an emergent ecocosmopolitanism based on dynamic global networks of information, communication, and culture.[30] Heise is correct in thinking that the old "idealistic" bioregional model is endangered, but it is not yet extinct. I suspect that in rural Nigeria or India, for example, such ecocosmopolitanism will not resonate strongly with its resident population. A "sense of planet" as opposed to an emergent sense of the local—or, as Lorna Lippard has put it, "the lure of the local"— is up against a space of biophysical sustenance forged within symbiotically connected biomes at local scale.[31] Break those fragile symbiotic bonds that still hold survival in situ, and all hell breaks loose.

The idea that "true patriots" might also protect and conserve that which is ecologically endemic and unique to a particular region is a possible bridge between those whose identity is strongly parochial and those who are sumbiosic. The term "patriots" has been used within the environmental movement to describe an endemic love of place and those willing to conserve and protect it.[32] The traditional term "conservative" been also used, to imply the conservation of energy and matter in the form of ecosystems that support life.[33] Within the environment movement, especially in the UK, critics of globalization have highlighted the nexus between identity and local-regional environments.[34] As local cultural and ecological diversity are removed from landscapes, so too is local identity.

Both local people and global people have their identity tested under stress. In the Middle East, under historical religious and political forces, plus crop and livestock failures due to anthropogenic climate warming exacerbating periodic drought, people have already moved and become refugees in

Europe and other parts of the world (those willing to accept them). Their movement not only obviously challenges their own identity, it challenges those who live in the regions they move to.

Ultimately, no matter how we think about ourselves, we are a biological species with relatively straightforward needs for food, water, shelter, and safety. Once our biological needs are put at risk as biophysical thresholds are crossed, humans will do anything to get access to these basic needs, including migrating and, if desperate, taking by force the necessities of life from other humans.

It seems we are back to the observation made by Margulis that while life in the biophysical world has been busy building symbiotic relationships over 3 million years, humanity has been extremely busy breaking them over the last three centuries. The symbiotic-busting form of global development we have committed to is a disaster for ecology, the climate, and ultimately for all biodiversity on the planet. It is also a disaster for human well-being and identity. Baby boomers and the old and new oligarchs of the planet are busy committing tierracide, while their offspring are engaging in global tourism and watching Netflix and "Match of the Day." As Bill Rees has passionately pondered:

> Why are we not collectively terrified or at least alarmed? If our best science suggests we are en route to systems collapse, why are collapse—and collapse avoidance—not the primary subjects of international political discourse? Why is the world community not engaged in vigorous debate of available initiatives and trans-national institutional mechanisms that could help restore equilibrium to the relationship between humans and the rest of nature?[35]

The answer to Bill Rees's question comes at least in part from the fact that people, the great majority of them, are not global citizens, informed and ready to be engaged in geo-eco-political and transnational discussion about imminent ecocollapse. There is no "global community," and people are busy at the local level avoiding the threats to life, such as starvation, war, rape, violence, and drug abuse. In other places people are trying to evade unemployment and underemployment due to economic rationalism, casualization, draconian governance and surveillance mechanisms, ubiquitous replacement of human labor by automation, and substitution by artificial "intelligence," all while worrying about being run over by a self-driving electric car. At the same time, prices for housing, rent, health care, energy, and food keep going up, and nation-states under economic rationalism continue to privatize all of the former "essential services," including, in some places, potable water. Survival in the metropolis is now almost as hard as survival in the village. Being homeless in the city might be far worse than being home-less in rural areas since in the village at least, your relatives might look after you and give you shelter.

Emergent Sumbioregional Identity

From what I have outlined above, human identity for all—males, females, LBGTQ people, and more—has become a place of dangerous contradic-tions, with globalization destroying the planet while locking people into psychic instability, toxic jobs, structural unemployment, addictive consump-tion, and deep inequality. Control and influence over our emotions in such a context has become a battleground.

Tribalism is built into our evolutionary origins and was once a vital re-quirement for our survival. Human kinship, tribalism, and territorial identity seem to be, from an understanding of our evolution as a species, universal human attributes.[36] While human kinship seems less relevant than tribalism and territory in contemporary society, the other two have not diminished in their influence over human behavior. With considerable trepidation, I sug-gest that tribe and territory may yet be once again important, as a form of emotional tribalism might be the only human social and cultural expression compatible with a Symbiocene world. The neotribalism of the defenders of "tradition" have a point. Their worldview is one that can be defended on neobioregional grounds that bring together the essential elements of hu-man culture, ecosystems, geology, and the biocomunen of biodiversity into some sort of functioning symbiotic consortium. The cosmopolitan alterna-tive takes us down the path to gigantism, and then ecocide, and ecocosmo-politanism increasingly looks like the last gasp of a privileged sector of the globalization juggernaut.

Given what I suggested above about the blending of cultural and bio-logical diversity, the strict boundaries of catchments and landforms of older forms of bioregionalism become less important than the overall sense of co-hesion that people have within a given geographical space. As argued above, the porosity of boundaries is reinforced by the concatenation of people and other biota from all over the world in places that were once relatively isolated.

We need to go beyond older conceptions of bioregionalism in order for the new emergent order of places to be understood, then made viable for habitation. The past, culturally and ecologically, lies hidden just below a new layer, the dominant strata of the Anthropocene. The older layers will need to be rediscovered, valued, and appreciated alongside the emergence of new but intimately related culture and ecology. New biomes will emerge in this process, and they will generate new forms of endemophilia in the people who live within them.

To give expression to this emergent idea, I define a "sumbioregion" as an identifiable biophysical and cultural geographical space where humans live together and engage in a common pursuit of the reestablishment and creation of new symbiotic interrelationships between humans, nonhuman organisms, and landscapes. More fine-tuned self-sufficiency within sumbioregions ipso facto means less global destruction. With more emotional "grounding" in the local, people become more secure. Terroirism just might also be a counter to terrorism and war. More democracy within regions will phase into sumbiocracy, as that which is inherent and endemic to a sumbioregion is perceived to be valuable and must be conserved. Confederations of sumbiocracies can form larger units of collaboration, and there is nothing stopping that process going global scale. The United Nations (UN) can be replaced by the United Sumbiocracies (US).

In essence, I am hopeful that a new form of human community, sumbio-tribal in its protection and love of place, can arise out of the current identity crisis of all the "techno-industrial" post-boomer generations and other people already emplaced at small scale all over the world. They will use their own inner distress, the scientific critique of homogeneous gigantism and its toxic legacy, climate chaos, plus sumbiophilia and tribalism, to create the Symbiocene. The inertia of recalcitrants becomes unblocked by positive Earth emotions based on the understanding of the symbiotic revolution in science and the overwhelming appeal to be a part of a mentally and physically strong Gen S.

There will undoubtedly be conflict in this transition period. I hope it will be minimal and nonviolent. Yet we also must be prepared for violence.

The Third and Last War: The Emotional War between Terraphthora and Terranascia

Warfare has been part of the human condition since recorded history. The origins of the word "war" have meanings connected to states of confusion and being "mixed-up." I can see the relevance of "war" at a time in history where, to end the biophysical and conceptual confusion, we need a shock to our current systems. Many humans, the terraphthorans, are now at war with the environment they inhabit. They are also fighting terranascient humans who they see as blocking their freedom to further exploit. There are also the terranascient humans who can see the destructive nature of the terraphthorans and oppose them with all their psychic and physical energy. There are now two huge tribes opposing one another in this discombobulated world, and their confrontation seems inevitable. It is an emotional war between Terranascians and Terraphthorans.

We have then, all of the makings of World War Three (WW3). Why not make it official? I declare WW3. This new global war will not be fought with guns, but will be a war where positive Earth emotions will have to directly confront negative Earth emotions. Terraphthorans are out to destroy themselves and the Earth, while Terranascians are out to nurture themselves and the Earth. The battle lines could not be clearer. A psychoterratic drama that has been unfolding over many decades must now tip over into open emotional warfare.[37]

The overwhelming justifications for WW3 are to reinstate symbiotic connections between the forms of life at specific places on this Earth and to reestablish human and ecosystem health. The Anthropocene represents "dysbiosis," or the breaking of vital bonds between symbiont species, causing ill health and possible death at the microbiome level.[38] Dysbiosis can also occur at all the other levels of life organization. For example, the loss of symbiont relationships and "mother trees" in forest ecosystems destroys their integrity and health. The use of "cides," the killers of life in agriculture, wipe out soil symbiosis and the biocomunen. Warming and acidifying oceans kill the symbiont relationships that build coral reefs. At micro, meso, and macro levels of life, it is the bonds between symbiont species that maintain and perpetuate life and its health, including that for humanity. Dysbiosis is death. Symbiosis is life.

As a consequence, the primary work of Gen S will be to identify and maintain the life bonds and create new ones. As sumbioregions succeed in this task, their combined efforts will rebuild ecosystem health at island, continental, and global scales. For all the generations produced after the boomers, now is their time in history to unite and fight for a cause that is without peer: themselves and their own future. The artistic tribes have already moved in this direction, and I am buoyed by the emergence of Symbiocene art in the last few years. Symbiocene principles form a coherent and unifying ideology. The idea of the Symbiocene and its positive Earth emotions give generously of directionality in life.[39] They both say go in the direction of increasing symbiotic unity and diversity.

With emergent Symbiocene identity, and the anger and emotional out-rage to confront those who are trying to deny them a viable future, all the Gens become part of a force majeure to be reckoned with. Here is a uni-versal enemy that must be opposed in a thousand different contexts. The larger home, the Earth, provides universal identity as part of life, but within different locations on the planet there must be unique, symbiotically created diversity. There will be unity in diversity. Identity in differences.

When the Gens see this universal motivation for protecting and conserv-ing the distinctive parts of the world, they unite in their diverse but similarly motivated place-based repair and creation. The creative aspect of the Sym-biocene revolution goes beyond sustainability, regenerative and even some aspects of "restorative" thinking because Gen S will have to intensively in-novate to create novel symbiotic flows and emergent endemism in place and culture.

The War: Violence, Intelligence, and Men

Physical violence should never be the first-choice response to intimidation. The *sumbios* in the Symbiocene is founded on "living together"; negotiation, reason, and persuasion are its ethical foundations. But if the "ghehds" of people (humans and nonhumans) are violated, there must be a fallback and response position. If the other side is using threatening tactics that cannot be ignored, a response is required that is proportionate to their threats and actions.

I suggest that in addition to the sheer numbers constituting Gen S, "green muscle" will be needed to act as bodyguards and their emotional protectors. Defense and protection can be provided by a frontline show of strength when the terraphthoric side is intent on violence and assault. Many species avoid actual conflict by showing each other their strengths in acts of blatant display. Male Black Swans do this by synchronized swimming while posing a mirror image of each other until the bravado of one breaks down. In such a way, in addition to throwing flowers at the enemy, strong counter-intimidation can prevent actual violence.

There are instances, however, where preparation for armed conflict is already in situ to resist the ongoing activity of violent terraphthoric forces. For example, there are places where armed people (men and women) are needed to protect wildlife from poachers. In Africa, the poachers want ivory from elephant tusks or rhino horns, as this substance commands huge prices in other parts of the world. Trained and armed protectors are all that stand between preservation or extinction of these species in the wild, as there has been a catastrophic decline in their numbers over the last fifty years.

It is a tragedy, but the assassination of courageous symbiomental protectors and leaders who would likely support the idea of the Symbiocene is now a frequent occurrence. In late 2017, 164 people had been murdered, in various parts of the world, as they tried to defend special places and creatures of this Earth.[40] Clearly, even though I am a pacifist, I see the need to protect the protectors of the Earth with defense and yet more force, as they become the homicide targets of those committing ecocide and tierracide. Defense of life, even if it requires violence, must be permissible in the Symbiocene. After all, some of our gut bacteria work symbiotically with us to perform vital bodily defense, killing deadly bacteria in return for our supply of nutriments for them.

While physical strength, bravado, and aggression are not solely in the possession of men, their biology and evolutionary development has given emphasis to these attributes in the past. A revival of such attributes can be used to defend the defenders of the Symbiocene when they are under assault. Green muscle men will be like male Black Swans and will use their numbers and attributes for counter-intimidation purposes and defense.

In some measure, it is redundant bravery and physical strength, once seen as male virtues, that explain a good deal of the frustration in contemporary men and their maleness. Once "alpha" male characteristics now have little relevance in the technological world. One of the consequences of misplaced muscle and "strength" is the appeal of tribalism and the solidarity gained between men who are goal directed and hardwired to be tribal. There is also an acquired emotional power, bravado, somewhat equivalent to that which may have once been gained by physical prowess and athleticism. That emotional energy is now directed at preserving some sort of über-male identity, one that can be easily linked to nationalism and other male groups that require authority, order, and enemies in order to thrive. It is about the exercise of power and domination over those who do not fit within their worldview, and the appearance of bravery and strength in conformity with other men.

The universal dark suit and tie of corporate conformity in the globalized world is replicated or replaced with a symbolic "uniform" (e.g., the tattoo) that conveys a distinctively different identity yet conformity within their own patch and tribe. I put the case that some of this aggressive emotional power could be redirected at the forces tearing things apart in the world. While not all ecowarriors are men, many young men are willing to be in the front line of those protecting the remaining life places on Earth. They risk their personal safety, fines, and imprisonment to be guardians. Their bravery and strength should be seen as virtuous.[41] However, *all* strong and dexterous men and women (and all other genders) who wish to be part of the protection of Symbiocene protectors, now have a purpose for their training, skill, and hard work.[42] Green muscle will entail dangerous work but is absolutely necessary if people leading the Symbiocene revolution need to be protected from murderers. If symbolic and symbiotic resistance and advocacy fail, then conventional warfare to protect life becomes necessary. The outcome of such a war is likely to be catastrophic for all life on Earth.

Gen S, Solastalgia, and Symbiocene Identity

The loss of identity tied to a sense of place was one of the defining characteristics of my concept of solastalgia.[43] The issue of "identity" was even in the subtitle of the first publication devoted to it. In that first publication, I argued that while solastalgia is a serious psychic condition, it could be alleviated. I suggested:

> The defeat of solastalgia and non-sustainability will require that all of our emotional, intellectual and practical efforts be redirected towards healing the rift that has occurred between ecosystem and human health, both broadly defined. In science, such a commitment might be manifest in the full redirection of scientific investment and effort to an ethically inspired and urgent practical response to the forces that are destroying ecosystem integrity and biodiversity. The need for an "ecological psychology" that re-establishes full human health (spiritual and physical) within total ecosystem health has been articulated by many leading thinkers worldwide. The full transdisciplinary idea of health involves the healing of solastalgia via cultural responses to degradation of the environment in the form of drama, art, dance and song at all scales of living from the bioregional to the global. The potential to restore unity in life and achieve genuine sustainability is a scientific, ethical, cultural and practical response to this ancient, ubiquitous but newly defined human illness.[44]

Given what I have written, I find it strange that some of the commentators on solastalgia have suggested that "the pain of solastalgia tends to be irreversible."[45] The futurist Bruce Sterling suggests that it delivers "permanent

mutilation" to its sufferers, and that rather than "dark euphoria," if you suc-
cumb to it, "it will do you in."[46] There is an element of truth in this, in that
when a place that has been relatively stable is desolated, it can never return
to its previous state. Technically, that is true of every place on Earth. As
the climate change denialists constantly remind us, change is occurring all
the time, and we can never reverse the thermodynamic arrow of time and
space. However, as a psychoterratic concept, solastalgia is a negative Earth
emotion, and emotions can be repaired and restored to a state similar to
that before the desolation. The key though, is the halting of the desolation
of place, and its restoration to a state that once again delivers solace and
sustenance to its "owners."

I think such a restoration project is similar to E. O. Wilson's idea of re-
serving half the surface of the Earth for the preservation of a "home" for
all the remaining biodiversity of the Earth. His book *Half Earth* presents
a coherent way of avoiding the endangerment, then extinction, of all the
megafauna of the Earth within the next one hundred years.[47]

We can speculate about "half brain," a bit like half Earth. As mentioned
in chapter 2, Gregory Bateson argued that our psychic homes, our seats of
consciousness, have been substantially rendered "insane" by the pollution
of the Anthropocene.[48] If we see the cumulative minds of the human species as "psychic Earth," then I hope we still have at least half of our positive
psychoterratic potential left. It is a glass half full situation.

Building on our residual intelligence and good emotions, Gen S has the
opportunity to preserve and protect what is left, then create from that base
a full-Earth/full-brain terranascient future. In an essay on solastalgia, published in 2012, I further argued thus:

With a new psychoterratic language to describe and "re-place" our emotions and feelings, powerful transformative forces are unleashed. Solastalgia
is fixated on the melancholic, but it is also a foundation for action that will
negate it. There is a positive side to psychoterratic classifications, one where
positive earth emotions and feelings such as biophilia, topophilia, ecophilia,
soliphilia and eutierria can be used to counter the negative and destructive.

There is a drama going on in our heads and hearts, where solastalgia can
be defeated by the simultaneous restoration and rehabilitation of mental,
cultural, and biophysical landscapes.

Now that solastalgia and other psychoterratic terms (both positive and
negative) are being established in the research literature and many forms
of popular culture, and as recognition of the damage that degraded and

desolated environments do to our mental health increases, it is possible that we can respond more effectively to simultaneously restore mental and ecosystem health.[49]

I am with E. O. Wilson in this fight to conserve and rehabilitate the biophysical. I want to add the positive, sumbiocentric psychoterratic as well, and I think that they are both worth fighting for. Indeed, you cannot have one without the other, since to replace the Anthropocene, anthropocentric thinking must be replaced with sumbiocentric thinking.

In the past, humans as a species have fought over all sorts of really trivial matters such as the borders of nation-states. Now, the stakes are nothing less than the complete restoration of the whole Earth, and the complete restoration and reintegration of the human psyche and body with the Earth at specific places. To end anthropocentrism and replace it with sumbiocentrism are outcomes worth fighting for. In the case of Generation Symbiocene, it is the fight for identity and the fight for a good life.

Androgynous Symbiotic Intelligence

We now live in a technological world driven by intelligence that is neither characteristically male nor female. There are now thousands of jobs in industry where intelligence and dexterity are the required attributes, not brute strength. In the growth sector of service jobs, intelligence and emotional clarity are the essential skills. Equal male and female intelligence, and equal emotional responsiveness, become the base for a new form of equality.

As Gen S emerges, young people will use their androgynous symbiotic intelligence to quickly map and assess the huge losses to Earth and mind under the carnage of the Anthropocene. We must have maps of the disappeared and the disappearing. Once that task is completed, the act of creating new symbiotic linkages becomes everyone's job, and it will keep people very busy.

The most urgent task for Gen S will be the protest against gigantism. Hacking into, in multiple ways, then destroying gigantism becomes one's patriotic duty. Everything, from how we think, to how we live, will have to be disrupted by the activists and hacktivists—then reconstructed as the Symbiocene. There can be no time for gambling and gaming; this genuinely productive work will be relentless and all-consuming. It will test the ability of the best minds in all the current Gens.

As the symbols of gigantism come crashing down, new forms of sumbio-regional enterprise employing people in real work will arise. The principles of the Symbiocene, as listed in chapter 4, will be applied to *everything* that humans do. Symbiomimicry will have plastic on the run in no time at all. We will be eating our cellulose-based food packaging and loving it. We shall enjoy living in houses made of bricks produced and constantly repaired by domestic fungi.

Gen S will be full of people with ideas founded on symbiotic intelligence, all tailored to their own region and its needs. Ideas that work anywhere can be shared, as patents that support the biocomunen cannot be privately owned with rights; they will be shared as ghehds. Further, as the old GNP goes down, new indicators, genuine sumbiosic indicators of gross sumbio-regional product (GSP), start going up. Cancerous growth is replaced by normal growth, or growth that assists maturity, metabolic activity, and re-production. The economy becomes both restorative of life and innovative of new forms of symbiotic activity. It will have a purpose and a clear direction.

Androgynous symbiotic intelligence now has the ability to see the future, not with perfect clarity, but with enough clarity to know that without a rapid global transition to the Symbiocene, the future is likely to be bleak if not catastrophic. This is new in human history, as we have never before had that kind of predictive capacity. It is not Nostradamus giving us this dire warn-ing. It is being delivered by some of the best minds on the planet. For the first time, however, the known catastrophe can be averted and a new positive future created, perhaps without violence.

The coming together of human emotional intelligence and the practical intelligence required to build a new foundation for human societies world-wide is the ultimate aim of the meme of the Symbiocene. In avoiding disas-ter, we simultaneously create a viable future. That is the fate and work of Generation Symbiocene.

I will elaborate further on the application of Symbiocene principles by focusing on what I would call the "hierarchy of sumbiosity," concentrating on some of the elements that we need at the foundation of such a hierarchy to nurture what we wish to have at the top.

Communication

In the Western world, the many positive attributes of the Gens, centered on high levels of education and technological sophistication, enable them to communicate with each and every Symbiocene-inspired movement worldwide. Communications systems, similar to the ones now being used in Western contexts, can also be transferred to developing contexts. This can be done without the need for corporate-owned monopoly technology and media, as already decentralized, cheaper, and simpler technologies enable "internets" to be created without huge scale or expense all over the world.[50] These networked but community-based and community-owned communications will be necessary to avoid the monopoly/oligopoly Internet giants that now control access to goods (Amazon), communications (Facebook and Twitter), and operating systems (Google, Apple, Microsoft). The Internet will return to one of its original purposes: to allow free and instant communication between common interest groups.

A global communications pathway will remain necessary, but it will have to be protected from commercial and surveillance interests by gateways that protect the Symbiocene principles and matters of human privacy and social justice. Commerce can also have its own internets, but they must be for business only. The old separation of church and state will be replaced by the firm separation of commerce and communication. All the post-boomer generations will easily make this transition, as many already know about the public domain software needed to make viable local and regional-scale communication. Those not yet in the IT world already enjoy such a separation.

If we want to forge and own our personal and regional identities, we will need to decouple from the artificial intelligence, algorithm-based drivers of an interconnected, but authoritarian, homogeneous, depersonalized, and insecure social media. At present, the monopolies' interests are not our interests, unless we happen to be shareholders in one of these companies. Increased shareholder value is all the algorithms deliver; they are not designed to support personal identity, the symbioment, or the common good. The recent revelation that cryptocurrency Bitcoin transactions could dominate Internet energy usage, and eventually all energy usage, within a few short years, drives the self-annihilating aspect of the Anthropocene point home.[51]

As I mentioned above, identity or personal insecurity is also linked to these emergent monopolies and oligopolies, and the dark web seems to get more and more access to mass data banks containing our personal information. The Internet giants, though, are as vulnerable to disruption as the big fossil fuel–based centralized power industries of last century.

Given the failure of all types of government in the capitalist world to break up web-based oligopolies and monopolies in the so-called free-market, it will not be long before tech-savvy Gens hack into the gigantism and start breaking it up. Once this happens, we are well on the way to the Symbiocene. Money wrenching will be a lot more fun (and safer) than monkey wrenching.[52]

Habitat

Cities, as they are currently configured, owe their existence to globalized gigantism, supported by the carbon-intensive global transport of goods by aircraft, ships, and trucks. Symbiocene planners agree with Bill Rees in seeing cities as "entropic black holes" failing in all superficial attempts at even pseudo-sustainability.[53] We will need to exit the megapolis and live in what we could call the "sumbiopolis."

I have no clear idea of what this kind of city will look like but would not rule out the "living city" of Frank Lloyd Wright's Broadacre City, complete with skyscrapers, forests, and factories, surrounded by agricultural fields; an assemblage he described as "patterns of cultivation mingling with good buildings."[54] I can also see Friedensreich Hundertwasser's organic, living design and architecture informing every aspect of the built symbioment.[55] Finally, architects and designers will freely create organic form and flow in their work. The era of the straight line and the box will cease, as engineers figure out how to replace Anthropocene concrete and steel (two of the highest emitters of carbon) with new materials that satisfy Symbiocene principles.

Energy

The baby boomers have used energy at profligate rates for the whole of their adult tenure on this Earth. They have set up and now depend on an energy-intensive economy. Gen S must dismantle centralized provision of energy and the power structures that operate and own it. Decentralized energy and technologies that can be scaled down to local levels will be a necessity right from the commencement of the Symbiocene. It is fortunate that such systems are already substantially in place. With the cost of locally sourced, clean, safe, and renewable energy now falling well below its fossil fuel and nuclear competition, this is a revolution that does not require a huge fight. The German *Energiewende* (energy revolution) will go global and be manifest in each locality, according to its own renewable energy potential. We must not forget, the energy we get from renewable sources is free!

Once smaller-scale energy technologies are up and running, they will be in everyday use by communities and households the world over. Clean, renewable electricity can also be gifted, shared, locally traded, or sold as peer-to-peer transactions around communities, using locally based currencies. We are well on the way to seeing renewable energy and decentralized politico-economic power structures destroy the business model of centralized energy power structures and the centralized banks and finance institutions that support them. They will also undermine the profligate "mining" energy needed to produce globally traded cryptocurrencies such as Bitcoin.

In addition, biophotovoltaic cells are reality, so meeting the strict Symbiocene principles for energy production is no longer science fiction. The final form of energy driving the Symbiocene will not be based on more of our current forms of renewable energy. They are the transition technologies. The next generation energy will be, for example, a cell that "is biodegradable and in the future could serve as a disposable solar panel and battery that can decompose in our composts or gardens."[56] Our light bulbs will shine like fireflies.

Symbiocene Food and Becoming a Sumbiovore

Gen S will appreciate more than any earlier generation that "you are what you eat." As we further investigate the human gut microbiome, it is clear that the relationship between the foods we consume and the characteristics of our gut bacterial communities is an intimate one. The nutrition we get from food depends on the health of our gut bacteria. Our total well-being, including our mental health, is intimately connected to the individual gut microbiome.

In the future, gene sequencing will deliver us our gut-flora profile and fine-tune our bodily needs and food to achieve optimal health. Regionally, and even individually, we will optimize our food and bodily health, all based on the coevolution of microbiomes with mesobiomes. This is not new, but it seems new, because we now have the technologies to make explicit that which was always implicit in our evolutionary past. With new knowledge, it becomes possible to repair the dysbiosis done to the symbiotic interconnections that in the past made our bodies and mesobiomes unhealthy.

For Gen S to eat ethically, and as a reflection of Symbiocene principles, they must become "sumbiovores."[57] Humans have already created agricultural systems that, in varying degrees, respect forms of life, and we maintain the fertility of places that produce food sustenance over the long term.

"Sumbioculture" enhances these existing "sustainable" forms of agriculture such as permaculture and organic and biodynamic farming.[58] They will all become consistent with the health of symbiotically unified ecosystems and will evolve into distinctive forms of sumbioculture within sumbioregions. With sumbioculture, we will finally be rid of the structural dysbiosis typical of industrial agriculture.

Sumbiosic food production will enhance the mutual interdependence between the nonliving foundations of life (geochemical systems) and all species as living beings. In doing so, such food production will conserve and maximize the symbiotically united living systems that constitute communities of life. Sumbiosic food celebrates the interconnectedness of life and rejects food-production systems that deplete the soil base, practice extensive monoculture, poison the food chain, render species extinct, introduce risky DNA into life that cannot be removed once introduced, and produce emissions that create global problems, such as anthropogenic climate change.

The consumption of sumbiosic food from sumbiosic agriculture will support human and ecosystem health and will nurture the maximum diversity of life consistent with the aim of producing food for humans (and other beings such as our companion animals). It is a tougher standard for food than "organic," as organic systems can currently be monocultures and have no connection to the foundations of life in bioregional contexts, nor do they have to explicitly nurture symbiotic practice.

If you eat sumbiosic food, you are a sumbiovore, and can be described as a sumbiotarian.[59] If you support the whole system of sumbiosic food production and consumption, within the broader category of sumbiosic human development, you are eating within the Symbiocene. Gen S will all be sumbiovores irrespective of whether they are omnivores, vegetarians, vegans, pescatarians, or carnivores.

Sumbiotarians understand that health is the outcome of a delicate and dynamic balance of the parts and the totality. They know that at the local scale, the symbiotic roles of bacteria, fungi, manure, compost, and bioturbation all assist in maintaining ecosystem health. At the larger scale, biogeochemical cycles, such as those that determine the balance of nitrogen, carbon, and water, are what determine the health of the Earth. When the total system is healthy, and human life is harmoniously integrated within it, then sumbioculture will have a ghedeistual element to it. The love of food is also the love of life. Cooks and chefs who understand this vital connection will be seen as alchemists of the ghedeist as they take the life energy from sumbiosically produced food and transfer it, via creative cooking, into the vitality of diners.

Health

Health is at the top of the Symbiocene hierarchy. It is the final outcome of all the foundational elements at the base, such as the provision of food, sumbiosic habitats, healthy ecosystems, and energy. Young people maturing now will enter a period in Earth history where reintegration of humans with the rest of life on this planet will take place. The positive role of health in leading this movement becomes crucial to full entry into the Symbiocene. Sumbiocentric thinking (focused on mutualistic, symbiotic connection between species at all scales) will allow thought to move between micro, meso, and macro levels of life systems and see the relationships that maintain, reinforce, or destroy good health.

As indicated above, Gen S will become the first generation of humans who will have knowledge gained from maps of their microbiome, mesobiome, and macrobiome in intimate detail. For example, they will, with respect to the microbiome, be able to see its relevance to human health. Gene sequencing opens a massive field of discoveries about what we are and who we are as living beings.[60] Ecological science continues to provide knowledge about interconnections in life and physical systems at meso levels with toxicology, for example, reaching levels of never-before achieved precision about exactly what is in our symbioment. Earth systems science provides global-scale knowledge, often delivered via satellite-based information and data collection, with scope and scale not possible in the past. Climate science is now a vital part of understanding our planetary-scale macrobiome and its health.

A sumbiological health perspective will allow all three levels to be interpreted and presented to people, in a way that depicts health as an outcome of the complexity of relationships between the three levels. It will also be future directed, as it will help humans move from the dis-ease of the Anthropocene (at all scales) to the optimal well-being of the Symbiocene. The new transdisciplinary discipline of sumbiology will enable collaboration of many types of perspectives, both scientific and cultural, in the study of health interconnections. The transdisciplinary will have its own discipline at last.[61]

Symbiocene health will provide a motive for young people to reintegrate their bodies, lives, and lifestyles to a grounded view of health. Moreover, since health is a shared property between trillions of organisms, within biomes at all levels, it offers the prospect of human sharing and collaboration in the maintenance and optimization of life and health for all.

Sex and Population

Many members of Gen S will find new ways to express their sexuality, and the contemporary ecosexual movement is also primarily an identity-based reconnection to the Earth. The symbiotic erotic is ghedeistual as well as sensual. As the writers of "Ecosex Manifesto" have put it:

> Ecosexual Is an Identity: For some of us being ecosexual is our primary (sexual) identity, whereas for others it is not. Ecosexuals can be GLBTQI, heterosexual, asexual, and / or Other. We invite and encourage ecosexuals to come out. We are everywhere. We are polymorphous and pollen-amorous. We educate people about ecosex culture, community and practices. We hold these truths to be self-evident: that we are all part of, not separate from, nature. Thus all sex is ecosex.[62]

The deliberate separation of forms of sexuality from reproduction, not simply by contraception, will be an important component of population control in the future and it will, for example, obviously run counter to the teachings of the Catholic Church. While the current pope, Francis, has been a strong supporter of a viable symbioment, he and the church have made no change to the Catholic Church's stance on birth control. As androgynous symbiotic intelligence takes over and presents powerful reasons for limiting reproduction, the old institutions preventing population control will collapse. New institutions celebrating sex and life will take the place of the old churches. In the Symbiocene, there is the possibility of honesty about the centrality of sex in people's lives and its vital connection to understanding love and life.

Population control in the Symbiocene then becomes easier to see as a sumbiosic solution to the twin problems of overconsumption and too many people. Many in Generation S will decide not to have children, while others will voluntarily keep reproduction to replacement level only (maximum of two children per couple). This commitment will override religious and other factors that generate population growth beyond the carrying capacity of sumbioregions and, ultimately, the planet. The sharing of the socialization of children becomes commonplace and these sumbioliterate children become vital to the future of the Symbiocene. As parents educating for all life, Gen S adults play a huge role in the education of the next generations. This work also becomes highly valued in the Symbiocene.

As they build their own unique identity, Gen S has the task of building the Symbiocene. Given the dire legacy of the Anthropocene, their work will have to be fast and furious. As I argued in chapter 4, every single artifact of the Anthropocene that does not meet foundational Symbiocene principles will have to be replaced, if that is possible, with an equivalent Symbiocene "sumbiofact."

Many toxic things (stuff) that fail that test must be abandoned. Once started, this conversion from parasitism and pollution to symbiotic and organic alternatives is unstoppable. Everything that we build and use must be able to be fully integrated into the rest of life. There can be no return to the cave, as this revolution takes the very best of existing, then new, technology with us into a more sumbioregional and tribal future. The competition between tribes to build and use new Symbiocene technologies will also be a huge motivating factor for Gen S to innovate. Symbiocene innovators and symbiotic inventors become the most highly rewarded people within new occupations.

There are many more elements to the Symbiocene hierarchy. I consider the development of the detail of all this work in applying the Symbiocene principles to be the task of those with the necessary specific skills and capabilities. Many of the people with these skills are baby boomers, and it is my hope that they will not be selfish, and will make gifts of their skills to Gen S. As they gift their wealth and skills to the cause of the Symbiocene, they too will seamlessly exit the Anthropocene. Their legacy, should they choose this direction late in life, will no longer be connected to wrecking the planet, but to rebuilding it alongside their own children and grandchildren. That will finally mean a life well lived and a good death.

The formation of Gen S will not be easy, but the turmoil will be worth enduring. The largest asset that Gen S will have is sheer numbers and mass soliphilia to achieve the Symbiocene will overwhelm all opposition. The transition from democracy to sumbiocracy will also not be easy.[63] Nation-states will not readily give up their power to sumbioregions and will try to use the police and military force to maintain their hold on power and privilege. However, members of the police and the military are also people, with their own desire for identity, and many also have children and relatives that they love. If a critical mass of people all over the world opposes the representatives of the Anthropocene, they will need no force behind them for protection. The 99 percent will confront the 1 percent in a successful global-scale reprise of the Occupy movement of 2011. Hopefully, the military and the police will simply become more "green muscle," opposing the residual terraphthoric in all its forms. They will transform into repairers and protectors.

The Symbiocene is a compelling idea for the simple reason that it presents an optimistic future for humans. It is based on nothing other than the way life works, as understood by the very best science, practical thinking, and the emotional makeup of humans. To symbiotically and lovingly merge with all life on this planet will liberate good Earth emotions in so many ways, from the artistic and technological, to the ghedeistual. Generation Symbiocene will be the first generation for many centuries to be able to look children in the eyes and tell them with love and openness that their future looks good. Those children will run outside and play in healthy air, and be full of symbioment fulfillment order, the very opposite of nature deficit disorder—their laughter, and the singing of birds, clear signals that all is well with the Earth.

Conclusion

Earth Emotions started as a psychoterratic journey with solastalgia and ended with Generation Symbiocene. It has not been a simple, linear progression, for the path from negative to positive Earth emotions is convoluted and difficult. The main reason for such difficulty is the sheer embeddedness of the Anthropocene, which has inexorably infiltrated every aspect of life in the early twenty-first century.

Naomi Klein and the Occupy movement were right in thinking that imminent ecological, economic, and climate collapse would ultimately "change everything."[1] Yet as I write at this point in time, nothing fundamental has occurred to stop the Anthropocene. If anything, its terraphthoric forces seem to be on the ascendancy. There have been many warnings about the suicidal path that humans are now on, and the most cogent warnings have come from the domain of science. The world's top climate, symbiomental, and earth system scientists tell us that cumulative limits to all forms of growth in the form of toxic pollution, species loss, and climate change will bring the current form of global civilization to the point of collapse. Futurists, "big picture" thinkers, and scientists over the last half century, from Alvin Toffler to Jared Diamond, William Ripple and his colleagues, and now the IPCC in 2018 have warned us about these trends, yet their warnings go largely unheeded.[2]

It seems that hard facts do not sway people's opinions and behavior about what to do next in the light of the need to change. Values may be a more powerful impetus of change, and there are clear examples of shifting baseline values in some countries with respect to women's rights, same-sex marriage, and racial divides. In Australia in 2017, for example, the government ran a plebiscite to test public opinion on support for same-sex marriage. The plebiscite won 62 percent support from the Australian public (79% of eligible voters voted in a noncompulsory vote), and the parliament then passed the legislation needed to bring same-sex marriage into law. While an interesting exercise in democracy, it troubled me that this issue could garner huge media attention, great public interest, and strong voter participation, yet climate chaos and degradation of the symbioment cannot. The reason why same-sex marriage won support is, in my view, because it revolved around the issue of love.

The past injustice prevented people who love each other from declaring that love in a public ceremony and having that bond of love recognized by law. When the legislation passed through the Houses of Parliament in Australia, the then prime minister and many others declared it a victory for "love." Love is a complex and powerful human emotion, one recognized as held in common by almost all humans, regardless of gender or sexual identity. With love on your side, it seems you can win important political battles.

I have made the case that, at the core of our problems, are human emotions that I call our Earth emotions. Our negative Earth emotions are awakened as responses when the particular objects of our love—our home, our place, our sumbioregion, our continent, and our Earth—are being violated. Negative Earth emotions flow from the realization that the mutually beneficial, symbiotic bonds between people and places are being broken by forces beyond their control, while positive Earth emotions flow when the relationship is strong and beautiful.

What I have attempted in this book is to place a form of love at the core of our Earth emotions. I appeal to my fellow humans to see their connection to the Earth as a loving relationship. While many before me have made the claim that the Earth or world is our lover, I have given the concept a new dimension, with love merging into the ghedeist, a secular spiritual reflection of the life force that enables love to proliferate.[3] I am also aware, however, that conventional love is almost worn out as a driver of conceptual change because, like so many concepts, it has been appropriated in the Anthropocene as yet another algorithm-based consumer signifier of materialism. I have supported love with the best of the knowledge gained from the symbiotic revolution in the life sciences.

The gradual success of solastalgia as a psychoterratic concept is an indication that this ghedeistual form of love is one that humans are increasingly feeling, as Anthropocene obscenities continue to attack the foundations of life. My aim in creating solastalgia was to pinpoint the emotional base of human-Earth relationships, to isolate the particular form of distress that is at the core of the shattering of these living bonds between people and place, and to set about repairing them by the use of positive emotions and values.

The story of soliphilia is one of local and regional people responding to Earth desolation by political and policy action. Such action will replace the negative with a repaired and revitalized place that once again delivers positive emotional sustenance. Emotional repair work is intimately tied to the biophysical restoration of degraded land. The point is that as humans involve themselves in the restoration project and heal damaged places, they also heal themselves.

As the symbiotic interconnections come back into the soil, the ecosystem, and the macrobiomes, the neural and emotional connections return the psyche to a form of health. What is new in this field is the discovery that many of the foundational forces are invisible to us. We were simply ignorant of them. The microterraphthoric and the microterranascient turn out to be hugely important to life at all scales so we must now pay them close attention. As all human cultures have known, to some extent, ecosystem health and human health are intimately interconnected. Create the right symbiotic relationships in particular places, life will be good. If all people in all places create these kinds of relationships, then the totality of life on Earth will be very good. Gaia is the net result of all of these actions, not its prime mover.

Positive Earth emotions are complex to analyze, and I have maintained that when our Earth relationship is at its best, it is unaffected by a sense of separation between the knower and the known. The immersion of eutierria is total, even if at times fear, as well as joy, fills the mind. The fact that, for the bulk of our time on this Earth, humans have not had to analyze or even be conscious of our positive relationships to the Earth accounts to some extent for our not naming or recording them in the languages that have coalesced to become the English language. They were taken for granted when the world gave generously and continuously of these connections.[4]

As the Anthropocene has peeled away the protective layers that held our positive Earth emotions in place, we have come to appreciate and value their role in our psychic health. I have illustrated this loss with my own example of witnessing the destruction of endemism and my endemic sense of place in Western Australia. As macroscale human forces have disturbed every aspect of life systems, the dialectic between positive and negative Earth emotions has been exposed in all places on Earth. I sense that we need to do much more work on positive Earth emotions (in all languages) to identify them and define them before they become lost or dead heritage. Rediscovering lost words for landscapes, natural objects, and natural processes is vitally important.[5] Yet there is more to be done. We must identify feelings and emotions that are being lost even if we have never named them in our past. It has become crucial to name them now. I have also argued that we desperately need new psychoterratic words for the new state of the world.

New work is also required because the microterranascient is even more important than the microterraphthoric, as it must prevail for life to continue and flourish. How do we have a loving, ghedeistual relationship with that which remains invisible to us? If our gut bacteria can affect our emotions, what can we call a good gut terranascient feeling?

At both macro- and microscopic levels, humans are facing issues, and incorporating discoveries, never before encountered in our history. It is not that the past history of ideas is no longer useful, but that we are being forced by the scale and speed of negative change to create new terranascient ways of responding to such novelty. The task is urgent because the Earth is a lover that cannot wait any longer, and the terraphthoric forces in the emotional war will try to hasten its demise as a home for complex life.

Earth emotions are only one part of this challenge, yet they are at the core of how we respond as a species. The balance between the positive and the negative is delicately poised right now, and there are also many who are in limbo, struck by ecoparalysis as a form of conceptual confusion. How do we think about an issue like global climate change, one that has never before been experienced by a globally interconnected humanity? How do we respond to the discovery, in the last thirty years, that we share the one lifetime with trillions of symbiotic bacteria inside and on us?

In the spirit of the concept of "sumbiology," the study of the life-supporting relationships at all scales, I invite scholars, thinkers, and artists of all kinds to take up the new concepts and ideas presented in *Earth Emotions* and critique, develop, and apply them further. They all must create, when the space is needed, new concepts for this emergent world. For my part, I can see a union of the sumbiohumanities with the sumbiosciences in the creation of the Symbiocene. What a great foundation for a university that would be.

I am a realist when it comes to human nature. I accept that we are far from perfect as a species, with our tribal, aggressive, egotistical, acquisitive, and warlike tendencies. Cultures all over the world have, however, engaged with the effort required to hold the terraphthoric emotions in their place. I can see that, in the near future, the conflict between humans over where we are rapidly heading may be massive. I have skirted around the problem of open warfare between nations and people over diminishing resources, degraded land, land reclaimed by the sea, and places ruined by human-enhanced catastrophe. Humans are now so technologically powerful that war is an endgame with no winners. I live in hope that the fear of MAD, mutually assured destruction, remains a global force preventing world war. If I am wrong on this issue, then in my mind, the future, or lack of one, is unthinkable.

A form of global peace with each other and nonhuman life is the only viable option for the future. To this end, I have given Gen S the task of taking our partially globally interconnected species in a different direction, into the Symbiocene. Perhaps surprising to some, this idea revolves around preserving our tribal nature as a species. It also might surprise some to see that I advocate a huge surge in a new form of economic and technological innovation and growth. Tribalism and Symbiocene growth respect both what we are as a species and the way the world works as a series of interconnected sumbioregions. Emergent hybrid cultures that embrace emergent endemism and new forms of symbiotically integrated landscapes will also fit comfortably within the global matrix of sumbioregions. The symbiotic revolution is a way of returning human life to the rest of life in a form that does not defy human nature or the symbioment.[6] That aim is neither idealist nor atavistic.

The Symbiocene is a meme that invites all humans to create a future where positive Earth emotions will prevail over the negative. To create the Symbiocene, we must destroy the Anthropocene and its parasitic and cancerous forms of growth. We simply cannot allow a small minority in one species to come to dominate all else. I fear that even E. O. Wilson's "Eremocene," or "Age of Loneliness," where there is nothing wild left on Earth, only people, agriculture, and domesticated animals, will not be accurate enough to describe the high point of the Anthropocene.[7]

I have also argued that the people who support leaving this Earth to seek extraterrestrial places to live are ignoring the ethical implications of our current apocalyptic direction. To repair and restore this Earth *before* mass human and nonhuman misery and death is now the highest good that I can think of. There is nothing else that comes even close. Children born today will be here on this Earth in 2100, all things being equal. As a step-father, father of two children, and the grandfather of a little girl, I am outraged that some humans prefer to invest in outer space rather than ensuring that this

Earth remains a haven for life in a heartless cosmos. *Earth Emotions* has been written to ensure that Lyra and her children will have a bountiful home. It has also been written so that the Malleefowl will lay her eggs when the constellation Lyra appears in the night sky.

The Symbiocene is my emotional path to a future for humanity that will be hugely creative, healthy, and beautiful. Every terranascient emotion we have will find an outlet in the Symbiocene. Solastalgia, along with all the other negative psychoterratic emotions, will be on the run by the 2070s. As Gen S builds the Symbiocene, solastalgia will become a distant memory, a word that the *Concise Oxford English Dictionary* will consider removing from its e-pages in 2100 because it has become redundant.[8] What a good day that will be. As I turn comfortably in my well-composted grave, I will consider that progression of events to be a job well done.

Glossary of Psychoterratic Terms

biocomunen The property of shared life in coexisting multiple organisms or holobionts.

ecoagnosy Lack of knowledge about, hence ignorance of, past ecological states.

endemophilia The particular love of the locally and regionally distinctive in the people of a place.

eutierria A positive and good feeling of oneness with the Earth and its life forces where the boundaries between self and the rest of nature are obliterated and a deep sense of peace and connectedness pervades consciousness.

ghedeist (the) The awareness of a spirit or force that holds all life together; a feeling of profound symbiotic interconnectedness in all life between the self and other beings (human and nonhuman) and their gathering together to live within shared Earth places and spaces. It is a secular feeling of intense affinity and sense of mutual empathy for other beings.

global dread The anticipation of an apocalyptic future state of the world that produces a mixture of terror and sadness in the sufferer for those who will exist in such a state.

mermerosity An anticipatory state of being worried about the possible passing of the familiar and its replacement by that which does not sit comfortably in one's sense of place.

meteoranxiety The anxiety that is felt in the face of the threat of the increasing frequency and severity of extreme weather events.

psychoterratic Emotions related to positively and negatively perceived and felt states of the Earth.

solastalgia The pain or distress caused by the loss or lack of solace and the sense of desolation connected to the present state of one's home and territory. It is the lived experience of negative environmental change. It is the homesickness you have when you are still at home.

soliphilia The giving of political commitment to the protection of loved home places at all scales, from the local to the global, from the forces of desolation.

sumbiocentric Taking into account the totality of life interests in the biosphere at all scales when making decisions about human needs.

sumbiocracy A form of government where humans govern for the symbiotic, mutually beneficial, or benign relationships in a sociobiological system at all scales. Sumbiocracy is rule for the Earth, by the Earth, so that we might all live together.

sumbiocriticism A form of social, cultural, and literary criticism that evaluates all forms of creative endeavor from the awareness of the degree of interconnectedness between human and nonhuman life and the way that all must live together in shared spaces.

sumbiography An account of the cumulative influences on a person's life, from childhood to adulthood, that have culminated in their values and attitudes toward the relationship between humans, other forms of life, and nature.

sumbiology The study of humans living together with the totality of life. Sumbiologists study life-supporting relationships between people, other biota, ecosystems, and biophysical systems at places from the local to global scales.

sumbiophilia The love of living together.

sumbiopolis A form of high-density social living that is based on design and construction offering a symbiotic habitat for all kinds of beings that live within it.

sumbiosic Those cumulative types of active and purposive relationships and attributes created by humans that enhance mutual interdependence and mutual benefit for all living beings so as to conserve and maximize a state of unity-in-diversity.

Symbiocene The era in Earth history that comes after the Anthropocene. The Symbiocene will be in evidence when there is no discernible impact of human activity on the planet other than the temporary remains of their teeth and bones. Everything that humans do will be integrated within the support systems of all life and will leave no trace.

symbioment A recognition that all life exists within living systems at various scales. There is no "outside" for life forms within the biosphere.

symbiomimicry A form of design and creativity where the symbiotic elements of life processes are replicated in all types of human enterprise. Humans do more than copy the form of life; they replicate the life processes that make the mutually beneficial associations between different life forms strong and healthy.

terrafurie The extreme anger unleashed within those who can clearly see the self-destructive tendencies in the current forms of industrial-technological society and feel they must protest and act to change its direction.

terranascia Earth creator.

terraphthora Earth destroyer.

tierracide Earth murder. The deliberate desolation of the Earth such that it can no longer support life and life-support processes.

tierratrauma Acute Earth-based existential trauma in the present.

topoaversion The feeling that you do not wish to return to a place that you once loved and enjoyed when you know that it has been irrevocably changed for the worse.

topophobia Fear of entering a biophysical place.

topopinia A deep longing to enter a place you have never been to.

Notes

Introduction

1. *Terraphthora*: Earth destroyer, from the Latin word *terra*, "earth," and the Greek word *phthorá*, "destruction." *Terranascia*: Earth creator, from *terra* and the Latin word *nasci*, "to be born."

2. Davidson and Wahlquist 2017.

3. Stanner 2009, 61.

4. Stanner 2009, 61.

5. Rose 1996, 29.

6. I had to find out more about this constellation after my son and his partner named their daughter Lyra. A further reference to Lyra is the central character in Philip Pullman's series *His Dark Materials*.

7. In this text I follow Australian convention, which is to capitalize the accepted common names of identifiable species.

8. Morieson 1999.

9. Morieson 1999.

10. Roberts and Mountford 1974, 96.

11. Roberts and Mountford 1974, 70.

12. Crutzen and Stoermer 2000.

13. McMichael, Woodward, and Muir 2017.

14. Jones 1969.

15. Such a statement is not meant to deny the proliferation of war after 1953 at a regional level in places like the Middle East, Africa, or Asia. I wish only to highlight the massive difference between global war and its impacts and regional wars with limited impact.

16. Pope Francis 2015.

1. A Sumbiography

1. Albrecht 2018a. This term is derived from the Greek words *sumbiosis* (companionship), *sumbion* (to live together), *sumbios* (living together), and, of course, *bio* (life) and *graphy* (from *graphein*, to write).

In science, the word "symbiosis" is used to indicate life forms living together, usually for mutual benefit. With "sumbiography," and all subsequent terms in this book that feature the root word "sumbios," I wish to indicate that these words are the result of the applications of the science of symbiosis to various aspects of social human endeavor. Hence there is now a family of concepts, created by me, with the key term "sumbios," living together, as its shared theme. These concepts will be introduced to the reader in chapter 4.

2. I use "Indigenous," with a capital "I," when referring respectfully and directly to Indigenous Australian people. This is the preference of many Aboriginal people because they wish to be identified as the original Australian Indigenous people, in contrast to other indigenous peoples around the world. When I speak of indigenous people more generally, I do not use the capital "I."

3. A "dinky-di" bushman is a man with genuine Australian bush skills.

4. The Silvereye is a small insectivorous bird with a ring of silver feathers around the eye.

5. Albrecht and Albrecht 1992.

6. Albrecht and Albrecht 1992.

7. I examine Tuan and topophilia in more detail in chapters 2 and 4.

8. Seddon 1997.

2. Solastalgia

1. Tuan 1974.

2. Maynard 2004.

3. Hofer 1934.

4. The Welsh have *hiraeth* and the Portuguese *saudade*, both of which have connotations of nostalgic homesickness.

5. See Albrecht 2005, 2011a, 2012b.

6. Casey 1993.

7. Casey 1993, 38.

8. Stanner 2009, 206–7.

9. See Albrecht et al. 2017.

10. Lloyd 1978, 148.

11. Durkheim 1951.

12. Camus 1973, 62.

13. See Lear 2006.

14. Rose 1996, 7.

15. Leopold (1949) 1989.

16. Mitchell 1946, 4.

17. Bateson 1973, 460.

18. Rapport and Whitford, 1999.

19. There is an affinity here to the contemporary idea of "dysbiosis," or a "bad way of living" that produces sickness and death.

20. Freud 1919.

21. Heidegger 1962.

22. Relph (1976) 2008.

23. My first attempt to conceptualize this new emotion started with "placealgia." However, "placealgia" was such an ugly word that it failed to excite my interest beyond looking for further "algia" (pain) words in English that possibly referred to human emotion and a concrete notion of place.

24. Albrecht 2005, 46. For the first publication that featured solastalgia, see Connor et al. 2003.

25. Some early interpreters of the new concept wanted to apply it to the changes that occur to places when people of different color, ethnicity, and religion move into their neighborhood. I incurred the wrath of right-wing white supremacists in the United States when I strongly refuted their claim that they were suffering from solastalgia when blacks and Latinos moved into their neighborhood and negatively changed their sense of place. Definitions are important!

26. MacSuibhne 2009.

27. Thompson 2007.

28. The solastalgia connections to suicide are complex. In 2018, the New York–based activist David Buckel committed suicide by immolation. He left notes indicating that his death was a personal protest against fossil fuel–driven climate change. Nancy Romer, a retired psychologist, wrote an obituary for Buckel that integrated ecoanxiety and solastalgia into the reasons for his suicide. See Romer 2018.

29. Khanna 2009.

30. Albrecht 2012a.

31. Smith and Woodward 2014; Watts 2015.

32. On wildfire, see Eisenman et al. 2015.

33. Pretty 2014.

34. Donaldson 2009.

35. Tschakert and Tutu 2010.

36. Cunsolo Willox et al. 2013.

37. MacKinnon 2016.

38. Yoder 2018.

39. McNamara and Westoby 2011.

40. Jetnil-Kijiner 2014.

41. Holthaus 2014.

42. Warsini et al. 2015.

43. Hendryx and Innes-Wimsatt 2013.

44. Albrecht 2010a and Gunnoe 2009.

45. Ryan 2012.

46. Klein 2014.

47. Louv 2008.

48. Louv 2011.

49. Albrecht 2012c, 8.

50. Daw 2012.

51. Tüür 2017.

52. Bekoff 2007.

53. See Albrecht 2009b, 2011a.

54. Garrard 2012.

55. Fredericksen 2012, 21.

56. Albrecht 2005; Connor et al. 2004; Higginbotham et al. 2006. All interviewee quotation arises from research undertaken by Connor, Higginbotham, and Albrecht.

57. See Connor et al. 2008; Connor 2016; Higginbotham et al. 2010.
58. Higginbotham et al. 2006.
59. Albrecht et al. 2007.
60. Albrecht et al. 2007.
61. Sartore et al. 2008.
62. Ellis and Albrecht 2017.
63. Submission to New South Wales Land and Environment Court, 2013.
64. Submission to New South Wales Land and Environment Court, 2013.
65. Submission to New South Wales Land and Environment Court, 2013.
66. Guilliatt 2012.
67. New South Wales Land and Environment Court 2013.
68. Kennedy 2016.
69. Munro 2012; Bevan 2012.
70. See Crutzen and Stoermer 2000.

3. The Psychoterratic in the Anthropocene

1. Albrecht et al. 2007.
2. *Hindustan Times* 2017.
3. Chevalier et al. 2012.
4. Pretty 2017; Louv 2014.
5. Pretty 2017.
6. Wilson 1992, 334.
7. McKibben 1990, 44–45.
8. See Krause 2017.
9. Baker (1967) 2017, 5.
10. See Lear 2006.
11. Wright 2013.
12. McCarthy 2006.
13. Foer 2009.
14. Hoskins 2014.
15. Macfarlane 2015.
16. Crutzen and Stoermer 2000.
17. Hamilton 2017, 48.
18. Larson 2014.
19. The late Stephen Hawking was also an advocate of leaving the Earth.
20. Albrecht 2005.
21. Albrecht 2012a.
22. In Klein 2014, 165.
23. Macfarlane 2016
24. Heneghan 2013; Heneghan 2018.
25. Heneghan 2013.
26. Louv 2008.
27. Louv 2008, 2.
28. Louv 2011, 62–63.
29. Sobel 1996.

30. Wallace-Wells 2017; Hamilton 2017.

31. Harris 2016.

32. Sobel 1996.

33. Wilson 1984.

34. Kellert and Wilson 1993, 86.

35. Kahn 1999.

36. Kahn, Severson, and Ruckert 2009, 41.

37. Pyle 1993.

38. Albrecht 2017b.

39. There is also a form of environmental anxiety linked to the anxiety felt when colonizing people are surrounded by an alien landscape that presents novel difficulties and/or where the colonizing act creates unintended consequences that threaten the chance of colonizing being successful. See Beattie 2011.

40. Leff 1990.

41. Dickinson 2008.

42. Albrecht 2011a.

43. See Helma et al. 2018; Clayton et al. 2017.

44. Doherty and Clayton 2011.

45. Gifford and Gifford 2016.

46. Verplanken and Roy 2013.

47. Albrecht 2016b.

48. Albrecht 2016a.

49. Lertzman 2015.

50. McKibben 1990, 195.

51. Albrecht 2014a.

52. Norgay 2004.

53. Lenzen et al. 2018.

54. Albrecht 2012b, 257.

55. Hemon 2013.

56. Sterling 2009.

57. In Lukas 1986.

58. Oxfam 2017.

59. Lertzman 2008.

60. Rees 2007.

61. Especially so for academics, who generally have a very large carbon footprint due to their international travel for research and conferences.

62. Albrecht 2012d.

63. Albrecht 2017a. (*Terra*, the Earth; Middle English *furie*, from the Latin *furia*, from *furere*, to rage.)

64. Reprinted in Baker (1967) 2017, 215.

65. Baker (1967) 2017, 215.

66. Fromm 2010.

67. Fromm 2010, 37.

68. See Cook et al. 1970; Weisberg 1970.

69. Higgins 2010.

70. Higgins 2017.

71. Rees 2010.

4. The Psychoterratic in the Symbiocene

1. See World Commission on Environment and Development 1987; Holling 2001; Walker and Salt 2006.

2. National Oceanic and Atmospheric Administration 2018.

3. Ceballosa, Ehrlich, and Dirzob 2017.

4. Holling 2001; Walker and Salt 2006.

5. Gallopín 2006.

6. Holling 2001; Ráez-Luna 2008.

7. Green 2017.

8. Albrecht 1994.

9. Monbiot 2017a.

10. Mitchell 1946, 37.

11. Carson 1962, 63–64.

12. Merchant 1980.

13. Clancy 2017.

14. Trappe 2005.

15. Margulis and Fester 1991.

16. Scofield and Margulis 2012, 232.

17. Hopper 2001.

18. Simard 2016, 249.

19. Stetka 2016.

20. Another surprising discovery is that microbiomes and fungal networks give generously to a long-appreciated characteristic of wine-growing regions. The famous French term "terroir," that distinctive quality in the soil, is now understood to be a product of the unique assemblage of microbiota of a given region and their influence on the chemical composition of a particular wine. Wine tasting may well be a test of a taster's ability to detect the signature of the unique microbiome of a wine sourced from a parcel of grapes from a single vineyard. The distinctiveness of regions worldwide is also, in part, a product of relatively undisturbed symbiotically unified microbiomes over extended time.

21. Margulis and Fester 1991.

22. Ingold 2006.

23. From *bios*, life, and *comunen*, to have common dealings with, Middle English, from Old French *communer*, to make common, share (from *commun*, common).

24. The full derivation is from the Greek *sumbiosis*, meaning companionship (from *sumbioun*, to live together, from *sumbios*, living together). In an earlier publication I used the term "symbionment" to match "environment," but have since abandoned it for a spelling with one less syllable.

25. Margulis and Sagan 1997, 195–96.

26. Albrecht 2016c.

27. Albrecht 2016d.

28. Meaning that I profess sumbiology, not that I hold the rank of Professor in a university system.

29. Albrecht 2014c.

30. Albrecht 2011b.

31. Wright and Nyberg 2016.

32. On the Dark Mountain Project, see Hine and Kingsnorth 2010.

33. See Hawken 2017.

34. Perhaps one of the initial important tasks of the discipline of sumbiology is to complete the critique of Anthropocene category mistakes. In particular, the concept of the "environment" must be removed from our thinking. However, another concept that must be critiqued is the concept of "ecology" and the related idea of an ecosystem. A sumbiology-based critique of ecology and an ecosystem must examine the idea that ecology is a prime example of the social construction of reductionist unreality from within Anthropocene science. "Ecology" has a root that it shares with "economics" (*oikos*: management of the household). That makes it difficult for it to be part of the Symbiocene because it opens the path for a form of systems thinking that puts artificial boundaries around non-bounded entities. One consequence of this abstraction is the ease with which ecology can be monetized by capitalism. Hence, the popularity within sustainability policy for ecosystem off-sets, the idea of natural capital and ecosystem services expressed as monetary value. It is a radical idea, but maybe ecology and ecological thinking have to be replaced.

35. For biomimicry, see Benyus 1997, and for symbiomimicry, Albrecht 2015b.

36. Albrecht 2013a.

37. Nature consists of everything both good and bad for humans. The asteroid that smashes into the Earth causing a great extinction event is part of nature. Life is that part of nature that creates and builds. Life is special, and it is qualitatively different from the cosmic drama that is the totality of raw nature.

38. "Sumbioism" is the term I now use for the collective and cumulative art and science of "living together" within the matrix of all life. A "sumbioist" is a person who reflects on, writes about, and lives by this philosophy.

39. From the Greek *sumbios,* meaning "living together," plus *cracy* meaning "rule." Rule for how we can all live together.

40. Albrecht 2015a.

41. Seed et al. 1988.

42. See MacIntyre 1984; Albrecht 1994.

43. Salleh 1984.

44. "Ghehd" has meanings linked to Old English and Germanic words such as "together," "to gather," and "good." See chapter 5.

45. United Nations 2016, 14.

46. Prescott and Logan 2017a.

47. Brown 2016; Fitzgerald 2016.

48. See Finegan 2016.

49. Giblett 2016, 2018.

50. Kropotkin (1901) 1987.

51. See Margulis 1998; Margulis and Sagan 1997; Bookchin 1991; Lovelock 1988; Haraway 2016; and Morton 2017.

52. Haraway 2016, 2.

53. Haraway 2016, 57.

54. Mitchell 1946, 28.

55. Perhaps also in defiance of her own husband, as Mitchell related in a letter: "Worse still my husband—just out of Changi—thought it [the book] stupidly uneconomically minded" (Mitchell, pers. comm., 1999). Mitchell would become famous, not for her environmental writing, but for her series of children's stories, which began with her novel *The Silver Brumby.* I hope that Mitchell will now be celebrated as a great sumbiomental writer as well.

56. Mitchell 1946, 138.

57. Mitchell 1946, 7.

58. Gammage 2012; Pascoe 2014.

59. Mitchell 1946, 139.

60. Mitchell 1946, 33.

61. See Nicholls 2014.

62. Elkin 1956, 200.

63. "Proto" meaning "first" not "primitive."

64. Stanner 2009, 64–65.

65. The idea that Aboriginals caused extinction of the megafauna in Australia is often proffered as proof that they did not have a benign relationship with their environment. Rather than acquiescence to the "noble savage" tradition, my reluctance to enter the great megafauna extinction debate is an outcome of my careful evaluation of the current science. Recent research in Australia indicates that "early humans to Australia lived alongside some of the megafauna for many thousands of years before the animals became extinct" (see Westaway, Olley, and Grun 2017). Climate change seems to be a more important factor in megafauna extinctions than human hunting and "overkill."

66. Based on the views of the English philosopher Thomas Hobbes (1588–1679).

67. Gammage 2012.

68. In chapter 6, "Generation Symbiocene," issues such as male violence and masculinity are discussed in the context of patriarchal-environmental destruction and ecocide. The idea of symbiosis as the basis of an androgynous symbiomental intelligence is also explored.

69. Mitchell 1946, 40 (my emphasis).

70. Mitchell 1946, 33.

71. Mitchell 1946, 33.

72. Mitchell 1946, 137 (my emphasis).

73. See Margulis 1992; Bateson 1973.

74. Mitchell 1946, 4.

75. Mitchell 1946, 23.

76. Scofield and Margulis 2012, 221 (my emphasis).

77. Davidson and Wahlquist 2017.

78. Schweitzer (1923) 1967, 11.

79. Fromm (1964) 2010, 43.

80. Fromm 1994, 101.

81. Sobel 1996, 5–6.

82. Wilson 1984, 1.

83. Margulis and Sagan 1997.

84. Wilson 2016, 122–23.

85. Wilson 2016, 212.

86. See David Orr's *Earth in Mind* for a detailed study of biophilia as a key to the future. Orr 2004.

87. Wilson 2016, 20.

88. Albrecht 2016e.

89. Albrecht 2014b.

90. Hauser 2007.

91. In Hauser 2007, 1.

92. Tuan 1974, 93.

93. Tuan 1974, 99.

94. See Morton 2017. In *Humankind* Morton puts solidarity right into the center of "leftist" symbiotic thinking.

95. Albrecht 2009a.

96. See Smith 2010; Worthy 2016.

97. MacDowell 2010.

98. Pers. comm. 2009.

99. Robinson n.d.

100. See Klein 2014.

101. Albrecht 2013b.

102. Garkaklis, Bradley, and Woller 1998.

103. Garkaklis, Bradley, and Woller. 1998

104. Relph (1976) 2008.

105. Relph (1976) 2008, 55.

106. In Broome 1982, 14.

107. Weldon 2014.

108. Albrecht 2010b.

109. von Humboldt 1995, xliii

5. Gaia and the Ghedeist

1. Jung 1964, 95.

2. Mitchell 1946, 140–41.

3. See Mitchell 1947, 1946, 1945.

4. Albrecht 1989, 212.

5. Lovelock 2006.

6. Margulis 1998, 118.

7. See Ruether 1992; Harding 2006.

8. Clark 1983, 195.

9. In Lovelock 2006, xv.

10. In Lovelock 2006, xv.

11. Albrecht 2016f.

12. The latest iteration of this idea is to be able to transfer human intelligence to machines, a process Elon Musk calls a "neural lace" that will help humans "achieve symbiosis with machines," a subset of a movement known as transhumanism. "Symbiosis" with machines is not symbiosis, as there is no living, mutualistic relationship between two or more beings.

13. Avatar Wiki n.d.

14. James Cameron Online n.d.

15. Piazza 2010.

16. Albrecht 2010c.

17. See, for example, Holtmeier 2013.

18. Taylor 2013.

19. An exercise in cultural criticism and ecocriticism would no doubt find many more issues to debate, as would criticism from feminist perspectives. As indicated above, I have restricted my comments to the psychoterratic, and they are not an endorsement of the film as a form of entertainment.

20. Fredericksen 2012, 21.

21. Mitchell 1946, 90.

22. Christie 2016, 16.

23. Casselman 2007.

24. It must be noted that pathogenic fungi also extend over large expanses of forest worldwide, killing many different kinds of trees, including the noble Ash and the mighty Jarrah. Phytophthora kills—it is a plant destroyer—but its rapid desolation of a Jarrah forest ecosystem can happen only because of human disturbance, anthropogenically introduced pathogens, and plant stress due to climate warming. The endemic forests were resilient for millennia because they evolved mutually beneficial symbiotic relationships with the endemic fungi and bacteria. Globalization and the great warming changed everything at the microbiome level.

25. Bassler 2007.

26. Hug et al. 2016.

27. Borgonie and Lau 2017.

28. Perhaps there is no such thing as an "ecosystem," and a concept that might be the last gasp of reductionism in biology.

29. Steiner (1923) 1998, prelude 2–3.

30. Steiner (1923) 2008, 125.

31. Neidjie 2002, 34.

32. Plato 1970, 200.

33. Hegel 1961, 304.

34. Hegel 1961, 304.

35. Hegel 1961, 308.

36. Hegel 1961, 278–79.

37. Hegel 1961, 290.

38. Fromm 2010, 43.

39. Fromm (1964) 2010, 41.

40. Battson 2015.

41. Kamenka 1962, 102.

42. Angier 2010.

43. Greer 2013, 113.

44. They might help strengthen the immune system on the basis of the old adage that "if it does not kill you, it makes you stronger." Our immune systems have coevolved with parasites, and to eliminate them might invite an immune system with nothing to do except cause trouble in the body. Hence a world without parasites could be a world with immune systems out of whack. Parasites also "take out" the weak and the elderly, another one of their roles in evolution.

45. Anderson, in Kamenka 1962.

46. Kamenka 1962, 103.

47. On fluminism, see Battson 2017.

48. Autopoiesis, or "self-making," is replaced by sympoiesis, or "making with others." See Haraway 2016, 58.

6. Generation Symbiocene

1. Baby boomers are generally defined as the generation born between 1945 and 1964.

2. Flannery 1994.

3. I will give boomers one last chance for repentance toward the conclusion of this book.

4. See Albrecht 2018b.

5. Holthaus 2017.

6. Geisler and Currens 2017.

7. See Ripple et al. 2017.

8. World Health Organization 2017.

9. Galeon 2017.

10. Ghosh 2017.

11. I admire people like Brian Cox and the late Stephen Hawking for their scientific contributions. I am also a supporter of space exploration and discovery. What I oppose is space exploitation and the idea that it is necessary to "save the Earth."

12. Jensen 2015.

13. The "Me Too" movement is connected to this meta-theme. Female identity asserted in opposition to male "violation" became an international social movement in 2018.

14. Quattrociocchi, Scala, and Sunstein 2016.

15. Generation Identity 2018.

16. In November 2017 a march in Poland attracted sixty thousand people in support of Polish nationalism. Many of them were also members of Generation Identity.

17. Albrecht et al. 2013.

18. Maynard and Haskins 2016.

19. Albrecht et al. 2009.

20. Murphy 2017.

21. See, for example, Thomashow 1999; Sale 2000; Berg 2009; Lynch, Glotfelty, and Armbruster 2012.

22. Sale 2000, 55.

23. Kahn 1999.

24. Barber 1992.

25. Barber 1992.

26. See, for example, Tuck, McKenzie, and McCoy 2014; Luisetti, Pickles, and Kaiser 2015.

27. Klein 2014.

28. Daly 2017.

29. Hauer 2017.

30. Heise 2008.

31. See Lippard 1997.

32. Lines 2006.

33. Scruton 2012.

34. See, for example, Macfarlane 2015; Kingsnorth 2008; Monbiot 2017b.

35. Rees 2017.

36. Anthropologists have long emphasized the tribal aspects of human nature and culture. In acknowledging it, I want to focus on our terranascient tribal emotions, while at the same time recognizing that terraphthoran aspects also exist.

37. I developed an early version of this theme in TEDxSydney (Albrecht 2010d).

38. From *dys*, "bad," and *biosis*, "way of living," which I understand to mean the opposite of a "good way of living" as manifest in healthy symbiotic life forms. See Prescott and Logan 2017b.

39. Albrecht 1998.

40. Watts and Vidal 2017.

41. One such young man, Ben Morrow, from Newcastle, Australia, fought a good fight to save the old growth forests of Tasmania in the mid-2000s. He was able to climb massive eucalyptus trees and "occupy" them, thus preventing them from being logged. One occupation involved the "Global Rescue Station," a platform suspended sixty-five meters off the ground

in the middle of the tall forest. Morrow died of cancer in 2009 at the age of thirty-four. In the Symbiocene, he and other defenders of the forests, like Dr. Bob Brown, will be remembered and honored as heroes.

42. I am not arguing that male physical strength is no longer used in the advanced industrial world to gain exploitative advantage over others. The issue of male violence toward women is testimony to the ongoing relevance of this kind of violence-driven inequality. However, actual violence (assault) is now almost universally condemned as both highly unethical and, in most countries, illegal. Other forms of symbolic violence from men toward woman are now also being openly questioned and resisted. The misogyny of men within advanced industrial countries is now crumbling under a relentless ethical and intellectual assault from women and men who value justice. Soon there will be no place to hide it. The intellectual equality of women with men is already the basis for universal humanity. We are *Homo sapiens* after all.

43. Albrecht 2005.

44. Albrecht 2005, 59.

45. Macfarlane 2016.

46. Sterling 2017.

47. Wilson 2016.

48. Bateson 1973.

49. Albrecht 2012a.

50. Internet Society 2017.

51. See Rogers 2017; de Vries 2018.

52. Earth First 2017.

53. Rees 1997.

54. Wright 1958, 198.

55. Hundertwasser 1997.

56. Zyga 2017.

57. Albrecht 2016h. From the Greek *sumbiosis*, from *sumbioun*, to live together, from *sumbios*, living together, and *vore*, meaning what one eats plus a member of a group who eats that way.

58. Albrecht 2016h.

59. Albrecht 2016h.

60. Only if it respects privacy of data as non-negotiable.

61. See Higginbotham, Albrecht, and Connor 2001.

62. Sprinkle and Stephens 2011.

63. Some aspects of sumbiocracy can be seen in the emergence of "new municipalism" in places like Barcelona in 2017–18.

Conclusion

1. Klein 2014.

2. Toffler 1975; Diamond 2005; Ripple et al. 2017; IPCC 2018.

3. See Macy 2003; Johnson 2005.

4. In nonliterate cultures, the naming of positive Earth emotions may well have occurred, yet the world is losing these concepts along with the cultures and the rest of their languages at a rapid rate.

5. The work of Robert Macfarlane is hugely significant here. See Macfarlane 2015.

6. I have not provided any detail on how quickly the economy of the Anthropocene falls to pieces during the dismantling of the capitalist economy. I think it fair enough to say that the change that changes everything will be profound. Yet as the Anthropocene contracts, the Symbiocene expands.

7. Wilson 2016, 20.

8. Not that it is there right now. It is, however, in many online dictionaries.

References

Albrecht, Glenn A. 1989. "Organicism: An Historical and Systematic Examination." PhD thesis, University of Newcastle.

Albrecht, Glenn A. 1994. "Ethics, Anarchy and Sustainable Development." *Anarchist Studies* 2 (2): 95–118.

Albrecht, Glenn A. 1998. "Ethics and Directionality in Nature." In *Social Ecology after Bookchin*, edited by Andrew Light, 92–113. New York: Guilford Press.

Albrecht, Glenn A. 2005. "Solastalgia: A New Concept in Human Health and Identity." *PAN (Philosophy, Activism, Nature)* 3:44–59.

Albrecht, Glenn A. 2009a. "Soliphilia: The Antidote to Solastalgia." *Healthearth* (blog). February 19. http://healthearth.blogspot.com.au/2009/02/soliphilia.html.

Albrecht, Glenn A. 2009b. "Animal Nostalgia and Solastalgia: The Animal Mind and Psychoterratic Distress." Presentation at the Minding Animals International Conference, Newcastle, Australia, July 17.

Albrecht, Glenn A. 2010a. "Solastalgia and the Creation of New Ways of Living." In *Nature and Culture: Rebuilding Lost Connection,* edited by Jules N. Pretty and Sarah Pilgrim, 217–34. London: Earthscan.

Albrecht, Glenn A. 2010b. "Eutierria." *Healthearth* (blog). June 2. http://healthearth.blogspot.com.au/2010/06/solastalgia-soliphilia-eutierria-and.html.

Albrecht, Glenn A. 2010c. "Avatar and Virtual Solastalgia." *Healthearth* (blog). January 14. http://healthearth.blogspot.com.au/2010/01/avatar-and-virtual-solastalgia.html.

Albrecht, Glenn A. 2010d. "Environmental Change, Distress and Human Emotion, Solastalgia." Filmed May 22, 2010. TEDxSydney. YouTube video, 16:12. https://www.youtube.com/watch?v=-GUGW8rOpLY.

Albrecht, Glenn A. 2011a. "Chronic Environmental Change and Mental Health." In *Climate Change and Human Well-Being: Global Challenges and Opportunities*, edited by Inka Weissbecker, 43–56. New York: Springer.

Albrecht, Glenn A. 2011b. "Symbiocene." *Healthearth* (blog). May 19. http://health earth.blogspot.com.au/2011/05/symbiocene.html.

Albrecht, Glenn A. 2012a. "The Age of Solastalgia." *The Conversation*. August 7. https://theconversation.com/the-age-of-solastalgia-8337.

Albrecht, Glenn A. 2012b. "Psychoterratic Conditions in a Scientific and Technological World." In *Ecopsychology: Science, Totems and the Technological Species*, edited by Peter H. Kahn Jr. and Patricia Hasbach, 241–64. Cambridge, MA: MIT Press.

Albrecht, Glenn A. 2012c. "Solastalgia." In *Life in Your Hands: Art from Solastalgia*, edited by Debbie Abraham, 8–9. Lake Macquarie, New South Wales: Lake Macquarie City Art Gallery. https://artgallery.lakemac.com.au/downloads/48B8E8A9AB58D2697 681DC7DF8CCBFA6F7572EEF.pdf.

Albrecht, Glenn A. 2012d. "Tierratrauma." *Healthearth* (blog). September 2. http://healthearth.blogspot.com.au/2012/09/tierratrauma.html?m=0.

Albrecht, Glenn A. 2013a. "Sumbiosity and Sumbiosic Development." *Healthearth* (blog). April 11. http://healthearth.blogspot.com/2013/04/using-greek-to-save-greeks-and-rest-of_11.html.

Albrecht, Glenn A. 2013b. "Endemophilia." *Healthearth* (blog). May 11. http://healthearth.blogspot.com.au/2013/05/endemophilia.html.

Albrecht, Glenn A. 2014a. "Topoaversion." *Healthearth* (blog). July 23. http://healthearth.blogspot.com.au/2012/11/another-psychoterratic-concept.html.

Albrecht, Glenn A. 2014b. "Tierraphilia." *Psychoterratica* (blog). April 17. https://glennaalbrecht.wordpress.com/2015/04/17/tierraphilia/.

Albrecht, Glenn A. 2014c. "Ecopsychology in 'the Symbiocene.'" *Ecopsychology* 6 (1): 58–59. doi:10.1089/eco.2013.0091.

Albrecht, Glenn A. 2015a. "Sumbiocracy." *Psychoterratica* (blog). September 2. https://glennaalbrecht.wordpress.com/2015/09/02/sumbiocracy/.

Albrecht, Glenn A. 2015b. "Symbiomimicry." *Psychoterratica* (blog). April 17. https://glennaalbrecht.wordpress.com/2015/04/17/symbiomimicry/.

Albrecht, Glenn A. 2016a. "Mermerosity and the New Mourning." *Psychoterratica* (blog). November 20. https://glennaalbrecht.wordpress.com/2016/11/20/mermerosity-and-the-new-mourning/.

Albrecht, Glenn A. 2016b. "Meteoranxiety." *Psychoterratica* (blog). July 20. https://glennaalbrecht.wordpress.com/2016/07/20/meteoranxiety/.

Albrecht, Glenn A. 2016c. "Sumbiocentric." *Psychoterratica* (blog). June 6. https://glennaalbrecht.wordpress.com/2016/06/06/sumbiocentric-and-sumbiocentrism/.

Albrecht, Glenn A. 2016d. "Sumbiology." *Psychoterratica* (blog). May 7. https://glennaalbrecht.wordpress.com/2016/05/07/sumbiology/.

Albrecht, Glenn A. 2016e. "Sumbiophilia." *Psychoterratica* (blog). February 21. https://glennaalbrecht.wordpress.com/2016/02/28/sumbiophilia-2/.

Albrecht, Glenn A. 2016f. "Sumbiocriticism." *Psychoterratica* (blog). January 22. https://glennaalbrecht.wordpress.com/2017/01/22/sumbiocriticism/.

Albrecht, Glenn. A. 2016g. "The Ghedeist." *Psychoterratica* (blog). June 6. https://glennaalbrecht.wordpress.com/2016/06/06/the-ghedeist/.

Albrecht, Glenn, A. 2016h. "On Becoming a Sumbiovore and a Sumbiotarian." *Psychoterratica* (blog). February 28. https://glennaalbrecht.wordpress.com/2016/02/28/becoming-a-sumbiovore-and-a-sumbiotarian/.

Albrecht, Glenn A. 2017a. "Terrafurie." *Psychoterratica* (blog). July 12. https://glennaalbrecht.wordpress.com/2017/07/12/terrafurie/.

Albrecht, Glenn A. 2017b. "Ecoagnosy." *Psychoterratica* (blog). July 27. https://glennaalbrecht.wordpress.com/2017/07/27/ecoagnosy/.

Albrecht, Glenn A. 2017c. "Solastalgia and the New Mourning." In *Mourning Nature: Hope at the Heart of Ecological Loss and Grief,* edited by Ashlee Cunsolo and Karen Landman, 292–315. Montreal: McGill-Queen's University Press.

Albrecht, Glenn A. 2018a "Sumbiography." *Psychoterratica* (blog). March 20. https://glennaalbrecht.wordpress.com/2018/03/20/sumbiography-from-the-greek-sumbios-living-together/.

Albrecht, Glenn A. 2018b. "Public Heritage in the Symbiocene." In *The Oxford Handbook of Public Heritage Theory and Practice,* edited by Angela M. Labrador and Neil Asher Silberman, 335–67. New York: Oxford University Press.

Albrecht, Glenn A., and Jillian Albrecht. 1992. *The Goulds in the Hunter Region of N.S.W. 1839–1840.* Naturae 2. Clayton: Centre for Bibliographical and Textual Studies, Monash University.

Albrecht, Glenn A., Cassandra Brooke, David Bennett, and Stephen Garnett. 2013. "The Ethics of Assisted Colonization in the Age of Anthropogenic Climate Change." *Journal of Agricultural and Environmental Ethics* 26 (4): 827–45.

Albrecht, Glenn A., Nick Higginbotham, Linda Connor, and Neville Ellis. 2017. "Social and Cultural Perspectives on Ecology and Health." In *International Encyclopedia of Public Health,* 2nd ed., 551–56. n.p.: Elsevier. https://doi.org/10.1016/B978-0-12-803678-5.00414-8.

Albrecht, Glenn A., Clive McMahon, David Bowman, and Cory Bradshaw. 2009. "Convergence of Culture, Ecology and Ethics: Management of Feral Swamp Buffalo in Northern Australia." *Journal of Agricultural and Environmental Ethics* 22 (4): 361–78.

Albrecht, Glenn A., Gina-Maree Sartore, Linda Connor, Nick Higginbotham, Sonia Freeman, Brian Kelly, Helen Stain, Anne Tonna, and Georgia Pollard. 2007. "Solastalgia: The Distress Caused by Environmental Change." Special supplement, *Australasian Psychiatry* 15:95–98.

Angier, Natalie. 2010. "Listening to Bacteria: By Studying Microbial Communications, Bonnie Bassler Has Come Up with New Ways to Treat Disease." *Smithsonian,* August 2010. https://www.smithsonianmag.com/science-nature/listening-to-bacteria-833979/.

Avatar Wiki. "Pandora." Accessed September 13, 2018. http://james-camerons-avatar. wikia.com/wiki/Pandora.

Baker, John Alec. (1967) 2017. *The Peregrine*. London: William Collins.

Barber, Benjamin. 1992. "Jihad vs. McWorld: The Two Axial Principles of Our Age—Tribalism and Globalism—Clash at Every Point except One: They May Both Be Threatening to Democracy." *Atlantic*, March 1992. https://www.theatlantic. com/magazine/archive/1992/03/jihad-vs-mcworld/303882/.

Bassler, Bonnie. 2007. "Bacteria Talk." NOVA scienceNOW, PBS. January 1, 2007. http://www.pbs.org/wgbh/nova/body/bassler-bacteria.html.

Bateson, Gregory. 1973. *Steps to an Ecology of Mind*. London: Granada Publishing Limited in Paladin Books.

Battson, Ginny. 2015. "Mycelium of the Forest Floor. And Love." *Seasonalight* (blog). October 12. https://seasonalight.com/2015/10/12/mycelium-of-the-forest-floor-and-love/.

Battson, Ginny. 2017. "Beavers Are Fluminists." *Zoomorphic* 9. October 9. http://zoomorphic.net/2017/10/beavers-are-fluminists/.

Beattie, James. 2011. *Empire and Environmental Anxiety: Health, Science, Art, and Conservation in South Asia and Australasia, 1800–1920*. Cambridge Imperial and Post-Colonial Studies Series. New York: Palgrave Macmillan.

Bekoff, Mark. 2007. *The Emotional Lives of Animals*. Novato, CA: New World Library.

Benyus, Janice. 1997. *Biomimicry: Innovation Inspired by Nature*. New York: Harper Collins.

Berg, Peter. 2009. *Envisioning Sustainability*. San Francisco: Subculture Books.

Bevan, Scott. 2012. *The Hunter*. Sydney: Harper and Collins.

Bookchin, Murray. 1991. *The Ecology of Freedom: The Emergence and Dissolution of Hierarchy*. Montreal: Black Rose Books.

Borger, Julian. 2017. "Disruption Games: Why Are Libertarians Lining Up with Autocrats to Undermine Democracy?" *Guardian*, November 19, 2017. https://www.theguardian.com/technology/2017/nov/19/trump-russia-fake-news-libertarians-autocrats-democracy.

Borgonie, Gaetan, and Maggie Lau. 2017. "Life Goes Deeper. The Earth Is Not a Solid Mass of Rock: Its Hot, Dark, Fractured Subsurface Is Home to Weird and Wonderful Life Forms." *Aeon*, November 24, 2017. https://aeon.co/essays/deep-beneath-the-earths-surface-life-is-weird-and-wonderful.

Broome, Richard. 1982. *Aboriginal Australians: Black Response to White Dominance 1788–1980*. Sydney: George Allen & Unwin.

Brown, Jenny. 2016. "The Hitchhiker's Guide to the Symbiocene." Jenny Brown. http://jennybrownjenny.com/collaborations/hitchhikers-guide-to-the-symbiocene/.

Camus, Albert. 1973. *The Rebel*. Harmondsworth: Penguin.

Carson, Rachel. 1962. *Silent Spring*. Harmondsworth: Penguin Books.

Casey, Edward S. 1993. *Getting Back into Place: Toward a Renewed Understanding of the Place-World*. Bloomington: Indiana University Press.

Casselman, Anne. 2007. "Strange but True: The Largest Organism on Earth Is a Fungus." *Scientific American*. https://www.scientificamerican.com/article/strange-but-true-largest-organism-is-fungus/.

Ceballosa, Gerardo, Paul R. Ehrlich, and Rodolfo Dirzob. 2017. "Biological Annihilation via the Ongoing Sixth Mass Extinction Signalled by Vertebrate Population Losses and Declines." *PNAS* 114 (30). doi:10.1073/pnas.1704949114.

Chevalier, Gaétan, Stephen T. Sinatra, James L. Oschman, Karol Sokal, and Pawel Sokal. 2012. "Earthing: Health Implications of Reconnecting the Human Body to the Earth's Surface Electrons." *Journal of Environmental and Public Health* (January 2012): 291541. doi:10.1155/2012/291541.

Christie, John. 2016. "Trees Are the Central Theme That Gives a Garden Strength." *Diggers Winter Garden*, Diggers Club Australia, 16–17.

Clancy, Kelly. 2017. "Survival of the Friendliest: It's Time to Give the Violent Metaphors of Evolution a Break." *Nautilus* 046. http://nautil.us/issue/46/balance/survival-of-the-friendliest.

Clark, Stephen. 1983. "Gaia and the Forms of Life." In *Environmental Philosophy*, edited by Arran Gare and Robert Elliot. St. Lucia: University of Queensland Press.

Clayton, S., C. M. Manning, K. Krygsman, and M. Speiser. 2017. *Mental Health and Our Changing Climate: Impacts, Implications, and Guidance*. Washington, DC: American Psychological Association and ecoAmerica.

Connor, Linda. 2016. *Climate Change and Anthropos: Planet, People and Places*. Abingdon: Routledge.

Connor, Linda, Glenn A. Albrecht, Nick Higginbotham, Sonia Freeman and Craig Dalton. 2003. "Environmental Change and Human Health: A Pilot Study in Upper Hunter Communities." In *The Airs, Waters, Places Transdisciplinary Conference on Ecosystem Health in Australia*, edited by Glenn Albrecht, 116–28. Newcastle NSW: The University of Newcastle.

Connor, Linda, Glenn A. Albrecht, Nick Higginbotham, Sonia Freeman, and Wayne Smith. 2004. "Environmental Change and Human Health in Upper Hunter Communities of New South Wales, Australia." *EcoHealth* 1, suppl. 2: 47–58.

Connor, Linda, Nick Higginbotham, Sonia Freeman, and Glenn A. Albrecht. 2008. "Watercourses and Discourses: Coalmining in the Upper Hunter Valley, New South Wales." *Oceania* 78 (1): 76–90.

Cook, Robert. E., William Haseltine, and Arthur W. Galston. 1970. "What Have We Done to Vietnam?" *New Republic,* January 10, 1970, 18–21.

Crutzen, Paul J., and Eugene F. Stoermer. 2000. "The 'Anthropocene.'" *International Geosphere-Biosphere Program Newsletter* 41:17–18.

Cunsolo Willox, Ashlee, Sherilee L. Harper, James D. Ford, Karen Landman, Karen Houle, and Victoria L. Edge. 2013. "The Rigolet Inuit Community Government. Climate Change and Mental Health: A Case Study from Rigolet, Nunatsiavut, Labrador, Canada." *Climate Change* 121:255–70.

Daly, Herman. 2017. "Contribution to 'Roundtable on Global Capitalism.'" Great Transition Initiative. June. http://www.greattransition.org/roundtable/global-capitalism-herman-daly.

Davidson, Helen, and Calla Wahlquist. 2017. "Australian Dig Finds Evidence of Aboriginal Habitation Up to 80,000 Years Ago." *Guardian*, July 19, 2017. https://www.theguardian.com/australia-news/2017/jul/19/dig-finds-evidence-of-aboriginal-habitation-up-to-80000-years-ago.

Daw, Robyn. 2012. "Life in Your Hands: Art from Solastalgia." Lake Macquarie City Art Gallery. https://artgallery.lakemac.com.au/downloads/48B8E8A9AB58D2697 681DC7DF8CCBFA6F7572EEF.pdf.

de Vries, Alex. 2018. "Bitcoin's Growing Energy Problem." *Joule* 2 (5): 801–9. doi:https:// doi.org/10.1016/j.joule.2018.04.016.

Diamond, Jared. 2005. *Collapse: How Societies Choose to Fail or Succeed*. New York: Viking.

Dickinson, Elizabeth. 2008. "The Green Issue." *New York Times Magazine*. April 20, 2008. http://www.nytimes.com/2008/04/20/magazine/20Live-a-t.html?pagewanted=3.

Diss, Kathryn. 2017. "Blockchain Technology Fuels Peer-to-Peer Solar Energy Trading in Perth Start-Up." ABC News. November 10. http://www.abc.net.au/ news/2017-10-11/blockchain-technology-fuels-peer-to-peer-energy-trading-start-up/9035616.

Doherty, T. S., and Susan Clayton. 2011. "The Psychological Impacts of Global Climate Change." *American Psychologist* 66:265–76.

Donaldson, Mike. 2009. *Burrup Rock Art: Ancient Aboriginal Rock Art of Burrup Peninsula and Dampier Archipelago Western Australia*. Mt. Lawley, Western Australia: Wildrocks Publications.

Durkheim, Émile. 1951. *Suicide: A Study in Sociology*. New York: The Free Press.

Earth First. 2017. "Monkeywrenching." http://earthfirstjournal.org/monkeywrenching/.

Eisenman, David, Sarah McCaffrey, Ian Donatello, and Grant Marshal. 2015. "An Ecosystems and Vulnerable Populations Perspective on Solastalgia and Psychological Distress after a Wildfire." *EcoHealth* 12 (4): 602–10.

Elkin, Adolphus Peter. 1956. *The Australian Aborigines: How to Understand Them*. Melbourne: Angus and Robertson.

Ellis, Neville, and Glenn A. Albrecht. 2017. "Climate Change Threats to Family Farmers' Sense of Place and Mental Wellbeing: A Case Study from the Western Australian Wheatbelt." *Social Science & Medicine* 175:161–68.

Finegan, Ann. 2016. "Solastalgia and Its Cure." *Artlink*, September 1, 2016. https:// www.artlink.com.au/articles/4523/solastalgia-and-its-cure/.

Fitzgerald, Cathy. 2016. "Entering the Symbiocene: A Transversal Ecosophy-Action Research Framework to Reverse Silent Spring." SlideShare. December 5. https:// www.slideshare.net/cathyart/entering-the-symbiocene-a-transversal-ecosophyac tion-research-framework-to-reverse-silent-spring.

Flannery, Tim. 1994. *The Future Eaters: An Ecological History of Australasian Lands and People*. Sydney: Reed Books.

Foer, Jonathan Safran. 2009. *Eating Animals*. London: Penguin Books.

Fredericksen, Don. 2012. "Solastalgia: When Home Is No Longer Home." *Spring: A Journal of Archetype and Culture* 88 (Winter): 11–22.

Freud, Sigmund. 1919. *The "Uncanny."* In *The Standard Edition of the Complete Psychological Works of Sigmund Freud*, vol. 17, *1917–1919: "An Infantile Neurosis" and Other Works*. London: Vintage Classics.

Fromm, Erich. 1994. *On Being Human*. New York: Continuum.

Fromm, Erich. (1964) 2010. *The Heart of Man: Its Genius for Good and Evil*. Riverdale, NY: American Mental Health Foundation Books.

Galeon, Dom. 2017. "Jeff Bezos: 'We Have to Go to Space to Save Earth.'" Futurism. November 6. https://futurism.com/jeff-bezos-space-save-earth/.

Gallopín, Gilberto C. 2006. "Linkages between Vulnerability, Resilience and Adaptive Capacity." *Global Environmental Change* 16:293–303.

Gammage, Bill. 2012. *The Biggest Estate on Earth, How Aborigines Made Australia*. Sydney: Allen and Unwin.

Garkaklis, Mark J., John S. Bradley, and Ron D. Woller. 1998. "The Effects of Woylie (*Bettongia pencillata*) Foraging on Soil Water Repellency and Water Infiltration in Heavy Textured Soils in Southwestern Australia." *Australian Journal of Ecology* 23:492–96.

Garrard, Greg. 2012. *Ecocriticism*. New York: Routledge.

Geisler, Charles, and Ben Currens. 2017. "Impediments to Inland Resettlement under Conditions of Accelerated Sea Level Rise." *Land Use Policy* 66:322–30. https://doi.org/10.1016/j.landusepol.2017.03.029.

Generation Identity. 2018. "Generation Identity: United Kingdom and Ireland." Accessed October 7 2018. https://www.generation-identity.org.uk/.

Ghosh, Pallab. 2017. "Hawking Urges Moon Landing to 'Elevate Humanity.'" BBC News. June 20. http://www.bbc.com/news/science-environment-40345048.

Giblett, Rod. 2016. *Cities and Wetlands: The Return of the Repressed in Nature and Culture*. London: Bloomsbury.

Giblett, Rod. 2018. *Environmental Humanities and Theologies: Ecoculture, Literature and the Bible*. London: Routledge.

Gifford, Eva, and Robert Gifford. 2016. "The Largely Unacknowledged Impact of Climate Change on Mental Health." *Bulletin of the Atomic Scientists* 72 (5): 292–97. doi:10.1080/00963402.2016.1216505.

Green, Miranda. 2017. "Trump Administration Swaps 'Climate Change' for 'Resilience.'" CNN. September 30. http://edition.cnn.com/2017/09/30/politics/resilience-climate-change/index.html.

Greer, Germaine. 2013. *White Beech: The Rainforest Years*. London: Bloomsbury.

Guilliatt, Richard. 2012. "What They're Saying Is That We Aren't Worth Anything." *Weekend Australian Magazine*, May 26–27, 2012, 18–22.

Gunnoe, Maria. 2009. "My Life Is on the Line." Patchwork Films: Award Winning Virginia Documentaries. http://www.patchworkfilms.com/mariamylife.htm.

Hamilton, Clive. 2017. *Defiant Earth: The Fate of Humans in the Anthropocene*. Crows Nest, New South Wales: Allen & Unwin.

Haraway, Donna. 2016. *Staying with the Trouble*. Durham, NC: Duke University Press.

Harding, Stephan. 2006. *Animate Earth: Science, Intuition and Gaia*. Totnes, UK: Green Books.

Harris, Alex. 2016. "Octopus in the Parking Garage Is Climate Change's Canary in the Coal Mine." *Miami Herald*, November 18, 2016. http://www.miamiherald.com/news/local/community/miami-dade/miami-beach/article115688508.html.

Hauer, Mathew E. 2017. "Migration Induced by Sea-Level Rise Could Reshape the US Population Landscape." *Nature Climate Change* 7:321–25. doi:10.1038/nclimate3271.

Hauser, K. 2007. *Shadow Sites: Photography, Archaeology and the British Landscape 1927–1955*. Oxford: Oxford University Press.

Hawken, Paul. 2017. *Drawdown: The Most Comprehensive Plan Ever Proposed to Reverse Global Warming*. New York: Penguin.

Hegel, Georg Wilhelm Friedrich. 1961. *On Christianity: Early Theological Writings by Friedrich Hegel*. Translated by Thomas Malcolm Knox. New York: Harper and Brothers.

Heidegger, Martin. 1962. *Being and Time*. Translated by John Macquarrie and Edward Robinson. New York: Harper & Row.

Heise, Ursula. 2008. *Sense of Place and Sense of Planet: The Environmental Imagination of the Global*. New York: Oxford University Press.

Helm, Sabrina V., Amanda Pollitt, Melissa A. Barnett, Melissa A. Curran, and Zelieann R. Craig. 2018. "Differentiating Environmental Concern in the Context of Psychological Adaption to Climate Change." *Global Environmental Change* 48:158–67.

Hemon, Aleksandar. 2013. *The Book of My Lives*. New York: Picador.

Hendryx, Michael S., and Kestrel A. Innes-Wimsatt. 2013. "Increased Risk of Depression for People Living in Coal Mining Areas of Central Appalachia." *Ecopsychology* 5 (3): 179–87.

Heneghan, Liam. 2013. "The Ecology of Pooh." *Aeon*, March 5, 2013. https://aeon.co/essays/can-we-ever-return-to-the-enchanted-forests-of-childhood.

Heneghan, Liam. 2018. *Beasts at Bedtime: Revealing the Environmental Wisdom in Children's Literature*. Chicago: Chicago University Press.

Higginbotham, Nick, Glenn Albrecht, and Linda Connor. 2001. *Health Social Science: A Transdisciplinary and Complexity Perspective*. South Melbourne: Oxford University Press.

Higginbotham, Nick, Linda Connor, Glenn A. Albrecht, Sonia Freeman, and Kingsley Agho. 2006. "Validation of an Environmental Distress Scale (EDS)." *EcoHealth* 3 (4): 245–54.

Higginbotham, Nick, Sonya Freeman, Linda Connor, and Glenn Albrecht. 2010. "Environmental Injustice and Air Pollution in Coal Affected Communities." *Health and Place* 16:259–66.

Higgins, Polly. 2010. *Eradicating Ecocide: Laws and Governance to Prevent the Destruction of Our Planet*. London: Shepheard-Walwyn.

Higgins, Polly. 2017. "State and Corporate Crime." Eradicating Ecocide. http://eradicatingecocide.com/our-earth/.

Hindustan Times. 2017. "Indian Cities May Face Deadly Heatwaves Due to Global Warming." April 23, 2017. http://www.hindustantimes.com/india-news/indian-cities-may-face-deadly-heatwaves-due-to-global-warming/story-saKEyNXUU2mi4SjMrhbKWJ.html.

Hine, Dougald, and Paul Kingsnorth, eds. 2010. *Dark Mountain*. Issue 1. The Dark Mountain Project.

Hofer, Johannes. 1934. "Medical Dissertation on Nostalgia." 1688. Translated by Carolyn Kiser Anspach. *Bulletin of the Institute of the History of Medicine* 2: 376–91.

Holling, Crawford Stanley. 2001. "Understanding the Complexity of Economic, Ecological, and Social Systems." *Ecosystems* 4:390–405.

Holthaus, Eric. 2014. "This Climate Change Poem Moved World Leaders to Tears Today." *Future Tense* (blog). *Slate*, September 23, 2014. http://www.slate.com/blogs/future_tense/2014/09/23/kathy_jetnil_kijiner_solastalgia_marshall_islander_s_poem_moves_u_n_climate.html.

Holthaus, Eric. 2017. "Ice Apocalypse: Rapid Collapse of Antarctic Glaciers Could Flood Coastal Cities by the End of This Century." *Grist Magazine*, November 21, 2017. https://grist.org/article/antarctica-doomsday-glaciers-could-flood-coastal-cities/.

Holtmeier, Matthew. 2013. "Post-Pandoran Depression or Na'vi Sympathy: Avatar, Affect, and Audience Reception." In *Avatar and Nature Spirituality*, edited by Bron Taylor, 83–94. Waterloo: Wilfred Laurier University Press.

Hopper, Steve. 2001. *Flora of Western Australia*. West Perth: Kings Park and Botanic Garden.

Hoskins, Tansy E. 2014. *Stitched Up: The Anti-Capitalist Book of Fashion*. London: Pluto Press.

Hug, Laura A., Brett J. Baker, Karthik Anantharaman, Christopher T. Brown, Alexander J. Probst, Cindy J. Castelle, Cristina N. Butterfield, Alex W. Hernsdorf, Yuki Amano, Kotaro Ise, Yohey Suzuki, Natasha Dudek, David A. Relman, Kari M. Finstad, Ronald Amundson, Brian C. Thomas, and Jillian F. Banfield. 2016. "A New View of the Tree of Life." *Nature Microbiology* 1:16048. doi:10.1038/nmicrobiol.2016.48.

Hundertwasser, Friedensreich. 1997. *Hundertwasser Architecture*. Cologne: Taschen.

Ingold, Tim. 2006. "Rethinking the Animate, Re-animating Thought." *Ethnos: Journal of Anthropology* 71 (1): 9–20.

Intergovernmental Panel on Climate Change (IPCC). 2018. "Global Warming of 1.5 °C" (special report). http://www.ipcc.ch/report/sr15/.

Internet Society. 2017. *Policy Brief: Spectrum Approaches for Community Networks*. October 10. https://www.internetsociety.org/policybriefs/spectrum/.

James Cameron Online. "Avatar." Accessed September 13, 2018. http://www.jamescamerononline.com/AvatarFAQ.htm.

Jensen, Derrick. 2015. "The Man Box and the Cult of Masculinity." *Deep Green Resistance News Service*. September 2. https://dgrnewsservice.org/civilization/patriarchy/rape-culture/derrick-jensen-the-man-box-and-the-cult-of-masculinity/.

Jetñil-Kijiner, Kathy. 2014. "Dear Matafele Peinam." Kathy Jetñil-Kijiner. October 24. https://www.kathyjetnilkijiner.com/united-nations-climate-summit-opening-ceremony-my-poem-to-my-daughter/.

Johnson, Trebbe. 2005. *The World Is a Waiting Lover: Desire and the Quest for the Beloved*. Novato, CA: New World Library.

Johnson, Trebbe. 2018. *Radical Joy for Hard Times: Finding Meaning and Making Beauty in Earth's Broken Places*. Berkeley, California: North Atlantic Books.

Jones, Rhys. 1969. "Fire-Stick Farming." *Australian Natural History* 16:224–28.

Jung, Carl. 1964. *Man and His Symbols*. London: Aldus Books.

Kahn, Peter H., Jr. 1999. *The Human Relationship with Nature: Development and Culture*. Cambridge, MA: MIT Press.

Kahn, Peter H., Jr., Rachel Severson, and Jolina Ruckert. 2009. "The Human Relation with Nature and Technological Nature." In *Current Directions in Psychological Science* 18 (1): 37–42. https://depts.washington.edu/hints/publications/Human_Relation_Technological_Nature.pdf.

Kamenka, Eugene. 1962. *The Ethical Foundations of Marxism*. London: Routledge and Kegan Paul.

Kellert, Stephen R., and Edward O. Wilson, eds. 1993. *The Biophilia Hypothesis*. Washington, DC: Island Press.

Kennedy, Amanda. 2016. "A Case of Place: Solastalgia Comes before the Court." *PAN: Philosophy Activism Nature* 12:23–33.

Khanna, Sanjay. 2009. "What Does Climate Change Do to Our Heads?" Culture Change. http://www.culturechange.org/cms/content/view/423/63/.

Kingsnorth, Paul. 2008. *Real England: The Battle against the Bland*. London: Portobello.

Klein, Naomi. 2014. *This Changes Everything: Capitalism vs. the Climate*. Melbourne: Allen Lane.

Krause, Bernie. 2017. "Mourning the Loss of Wild Soundscapes: A Rationale for Context When Experiencing Natural Sound." In *Mourning Nature: Hope at the Heart of Ecological Loss and Grief,* edited by Ashlee Cunsolo Willox and Karen Landman, 27–38. Montreal: McGill-Queen's University Press.

Kropotkin, Peter. (1901) 1987. *Mutual Aid as a Factor in Evolution*. London: Freedom Press.

Larson, Dean. 2014. "Mining Asteroids and Exploiting the New Space Economy." *Wall Street Journal,* August 21, 2014. https://www.wsj.com/articles/dean-larson-mining-asteroids-and-exploiting-the-new-space-economy-1408662987.

Lear, Jonathan. 2006. *Radical Hope: Ethics in the Face of Cultural Devastation*. Cambridge, Massachusetts: Harvard University Press.

Leff, Lisa. 1990. "Ecology Carries Clout in Anne Arundel." *Washington Post,* August 5, 1990. https://www.washingtonpost.com/archive/local/1990/08/05/ecology-carries-clout-in-anne-arundel/a01f0325-e1bf-4f25-b180-65bc4540ef0c/?utm_term=.04f611a3749d.

Lenzen, Manfred, Ya-Yen Sun, Futu Faturay, Yuan-Peng Ting, Arne Geschke, and Arunima Malik. 2018. "The Carbon Footprint of Global Tourism." *Nature Climate Change* 8:522–28. doi:10.1038/s41558-018-0141-x.

Leopold, Aldo. (1949) 1989. *A Sand County Almanac*. New York: Oxford University Press.

Lertzman, Renee. 2008. "The Myth of Apathy." Ecologist. June 19. http://www.theecologist.org/blogs_and_comments/commentators/other_comments/269433/the_myth_of_apathy.html.

Lertzman, Renee. 2015. *Environmental Melancholia: Psychoanalytic Dimensions of Engagement*. New York: Routledge.

Lines, William J. 2006. *Patriots: Defending Australia's Natural Heritage*. St. Lucia: University of Queensland Press.

Lippard, Lucy R. 1997. *The Lure of the Local: Senses of Place in a Multicentered Society*. New York: The New Press.

Lloyd, Geoffrey E.R., ed. 1978. *Hippocratic Writings*. Translated by John Chadwick and William Neville Mann. New York: Penguin.

Louv, Richard. 2008. *Last Child in the Woods: Saving Our Children from Nature-Deficit Disorder*. Chapel Hill, NC: Algonquin Books of Chapel Hill.

Louv, Richard. 2011. *The Nature Principle: Human Restoration and the End of Nature-Deficit Disorder*. Chapel Hill, NC: Algonquin Books of Chapel Hill.

Louv, Richard. 2014. "Ten Reasons Why We Need More Contact with Nature." *Guardian*, February 13, 2014. https://www.theguardian.com/commentisfree/2014/feb/13/10-reasons-why-we-need-more-contact-with-nature.

Lovelock, James. 1988. *The Ages of Gaia: A Biography of our Living Earth*. Oxford: Oxford University Press.

Lovelock, James. 2006. *The Revenge of Gaia*. London: Penguin.

Luisetti, Frederico, John Pickles, and Wilson Kaiser, eds. 2015. *The Anomie of the Earth: Philosophy, Politics, and Autonomy in Europe and the Americas*. Durham, NC: Duke University Press.

Lukas, J. Anthony. 1986. "The Rapture and the Bomb." *New York Times*, June 8, 1986. http://www.nytimes.com/1986/06/08/books/the-rapture-and-the-bomb.html?pagewanted=all.

Lynch, Tom, Cheryll Glotfelty, and Karla Armbruster, eds. 2012. *The Bioregional Imagination: Literature, Ecology, and Place*. Athens: University of Georgia Press.

MacDowell, Kate. 2010. "Soliphilia." Kate MacDowell. http://katemacdowell.com/soliphilia.html.

Macfarlane, Robert. 2015. *Landmarks*. London: Hamish Hamilton.

Macfarlane, Robert. 2016. "Generation Anthropocene: How Humans Have Altered the Planet Forever." *Guardian*, April 1, 2016. https://www.theguardian.com/books/2016/apr/01/generation-anthropocene-altered-planet-for-ever.

MacIntyre, Alasdair. 1984. *After Virtue; A Study in Moral Theory*. 2nd ed. Notre Dame, IN: University of Notre Dame Press.

MacKinnon, James B. 2016. "What Happens to a Town's Cultural Identity as Its Namesake Glacier Melts?" *Smithsonian Magazine*, February 16, 2016. http://www.smithsonianmag.com/science-nature/what-happens-town-cultural-identity-namesake-glacier-melts-180958140/.

MacSuibhne, Seamus. 2009. "What Makes a New 'Mental Illness'? The Cases of Solastalgia and Hubris Syndrome." *Cosmos and History: Journal of Natural and Social Philosophy* 5 (2): 210–25.

Macy, Joanna. 2003. *World as Lover, World as Self: Courage for Global Justice and Ecological Renewal*. Revised edition. Berkeley: Parallax Press.

Margulis, Lynn. 1992. *Symbiosis in Cell Evolution: Microbial Communities in the Archean and Proterozoic Eons*. New York: W.H. Freeman.

Margulis, Lynn. 1998. *Symbiotic Planet: A New Look at Evolution*. New York: Basic Books.

Margulis, Lynn, and René Fester, eds. 1991. *Symbiosis as a Source of Evolutionary Innovation: Speciation and Morphogenesis*. Cambridge, MA: MIT Press.

Margulis, Lynn, and Dorion Sagan. 1997. *Microcosmos: Four Billion Years of Evolution from Our Microbial Ancestors*. Berkeley: University of California Press.

Maynard, John, ed. 2004. *Awabakal Word Finder and Dreaming Stories Companion*. Southport, Queensland: Keeaira Press.

Maynard, John, and Victoria Haskins. 2016. *Living with the Locals: Early Europeans' Experience of Indigenous Life*. Canberra: NLA Publishing.

McCarthy, Cormac. 2006. *The Road*. London: Picador.

McKibben, Bill. 1990. *The End of Nature*. London: Penguin.

McMichael, Anthony. 2017. *Climate Change and the Health of Nations: Famines, Fevers, and the Fate of Populations*. With Alistair Woodward and Cameron Muir. Oxford: Oxford University Press.

McNamara, Karen E., and Ross Westoby. 2011. "Local Knowledge and Climate Change Adaptation on Erub Island, Torres Strait." *Local Environment: The International Journal of Justice and Sustainability* 16 (9): 887–901.

Merchant, Carolyn. 1980. *The Death of Nature: Women, Ecology and the Scientific Revolution*. San Francisco: Harper and Row.

Mitchell, Elyne. 1945. *Speak to the Earth*. Sydney: Angus and Robertson.

Mitchell, Elyne. 1946. *Soil and Civilization*. Sydney: Halstead Press.

Mitchell, Elyne. 1947. *Images in Water*. Sydney: Angus and Robertson.

Monbiot, George. 2017a. "How Do We Get Out of This Mess?" *Guardian*, September 9, 2017. https://www.theguardian.com/books/2017/sep/09/george-monbiot-how-de-we-get-out-of-this-mess.

Monbiot, George. 2017b. *Out of the Wreckage: A New Politics for an Age of Crisis*. London: Verso Books.

Morieson, John. 1999. "Neilloan." Victorian Malleefowl Recovery Group. http://www.malleefowlvictoria.org.au/aboriginalAstronomy.html.

Morton, Tim. 2017. *Humankind: Solidarity with Nonhuman People*. London: Verso.

Munro, Sharon. 2012. *Rich Land, Wasteland: How Coal is Killing Australia*. Sydney: Macmillan.

Murphy, Katy. 2017. "Calexit, Again: The Latest Campaign for California's Independence Has Some Ideas for the U.S. Constitution." *Mercury News*, October 17, 2017. https://www.mercurynews.com/2017/08/17/calexit-again-the-latest-campaign-for-californias-independence-has-some-ideas-for-the-u-s-constitution/.

National Oceanic and Atmospheric Administration. 2018. "Trends in Atmospheric Carbon Dioxide." https://www.esrl.noaa.gov/gmd/ccgg/trends/.

Neidjie, Bill. 2002. *Gagudju Man: Bill Neidjie*. Marleston, South Australia: JB Books.

New South Wales Land and Environment Court. 2013. https://www.caselaw.nsw.gov.au/decision/54a639943004de94513da836.

Nicholls, Christine Judith. 2014. "'Dreamtime' and 'The Dreaming': An Introduction." *The Conversation*. January 23. https://theconversation.com/dreamtime-and-the-dreaming-an-introduction-20833.

Norgay, Jamling Tenzing. 2004. "Mountains as an Existential Resource, Expression in Religion, Environment and Culture." *Ambio* 13:56–57. http://www.jstor.org/stable/25094589.

Orr, David. 2004. *Earth in Mind: On Education, Environment, and the Human Prospect*. Washington, DC: Island Press.

Oxfam. 2017. "An Economy for the 99%." Oxfam. January. https://www.oxfam.org.au/wp-content/uploads/2017/01/An-economy-for-99-percent.pdf.

Pascoe, Bruce. 2014. *Dark Emu Black Seeds: Agriculture or Accident?* Broome, Western Australia: Magabala Books.

Piazza, Jo. 2010. "Audiences Experience 'Avatar' Blues." CNN Entertainment. January 11. http://edition.cnn.com/2010/SHOWBIZ/Movies/01/11/avatar.movie.blues/index.html.

Plato. 1970. *Symposium.* Translated by Benjamin Jowett. London: Sphere Books.

Pope Francis. 2015. *Laudato Si: On Care for Our Common Home.* Rome: The Vatican.

Prescott, Susan L., and Alan C. Logan. 2017a. "Down to Earth: Planetary Health and Biophilosophy in the Symbiocene Epoch." *Challenges* 8 (2): 19. doi:10.3390/challe8020019.

Prescott, Susan L., and Alan C. Logan. 2017b. *The Secret Life of Your Microbiome: Why Nature and Biodiversity Are Essential to Health and Happiness.* Gabriola Island, British Columbia: New Society.

Pretty, Jules. 2014. *The Edge of Extinction: Travels with Enduring People in Vanishing Lands.* Ithaca, NY: Cornell University Press.

Pretty, Jules. 2017. "Manifesto for the Green Mind." *Resurgence and Ecologist,* March/April 2017. https://www.resurgence.org/magazine/article4826-manifesto-for-the-green-mind.html.

Pullman, Philip. (1995) 2007. *The Northern Lights: His Dark Materials.* London: Scholastic UK.

Pyle, Robert Michael. 1993. The Thunder Tree: Lessons from an Urban Wildland. Boston: Houghton Mifflin.

Quattrociocchi, Walter, Antonio Scala, and Cass R. Sunstein. "Echo Chambers on Facebook." Abstract. June 13. https://ssrn.com/abstract=2795110.

Ráez-Luna, Ernesto. 2008. "Third World Inequity, Critical Political Economy, and the Ecosystem Approach." In *The Ecosystem Approach: Complexity, Uncertainty, and Managing for Sustainability,* edited by David Waltner-Toews, James J. Kay, and Nina-Marie E. Lister, 323–34. New York: Columbia University Press.

Rapport, David, and Walter G. Whitford. 1999. "How Ecosystems Respond to Stress." *BioScience* 49 (3): 193–203.

Rees, William E. 1997. "Is 'Sustainable City' an Oxymoron?" *Local Environment* 2 (3): 303–10.

Rees, William E. 2007. "Global Change, Eco-Apartheid and Population." Presentation in Ottawa, Ontario, November 5, 2007. SlidePlayer. http://slideplayer.com/slide/8266430/.

Rees, William E. 2010. "What's Blocking Sustainability? Human Nature, Cognition, and Denial." *Sustainability: Science, Practice, and Policy* 6 (2): 13–25.

Rees, William E. 2017. "What, Me Worry? Humans Are Blind to Imminent Environmental Collapse." *Tyee,* November 16, 2017. https://thetyee.ca/Opinion/2017/11/16/humans-blind-imminent-environmental-collapse/.

Relph, Edward. (1976) 2008. *Place and Placelessness.* London: Pion.

Ripple, William J., Christopher Wolf, Thomas M. Newsome, Mauro Galetti, Mohammed Alamgir, Eileen Crist, Mahmoud I. Mahmoud, and William F. Laurance. 2017. "World Scientists' Warning to Humanity: A Second Notice." *BioScience* 67 (12): 1026–28. https://doi.org/10.1093/biosci/bix125.

Roberts, Ainslie, and Charles P. Mountford. 1974. *The Dreamtime Book*. Sydney: Reader's Digest & Rigby.

Robinson, Jancis. n.d. "Western Australia." Jancis Robinson. Accessed September 13, 2018. https://www.jancisrobinson.com/learn/wine-regions/australia/western-australia.

Rogers, Adam. 2017. "The Hard Math behind Bitcoin's Global Warming Problem." *Wired*, December 15, 2017. https://www.wired.com/story/bitcoin-global-warming/?mbid=social_fb.

Romer, Nancy. 2018. "Eco-Anxiety: A Meditation on David Buckel's Life and Death." *Indypendent*, May 8, 2018. https://indypendent.org/2018/05/eco-anxiety-a-meditation-on-david-buckels-life-and-death/.

Rose, Deborah Bird. 1996. *Nourishing Terrains: Australian Aboriginal Views of Landscape and Wilderness*. Canberra: Australian Heritage Commission.

Ruether, Rosemary Radford. 1992. *Gaia and God: An Ecofeminist Theology of Earth Healing*. New York: Harper San Francisco.

Ryan, John Charles. 2012. *Green Sense: The Aesthetics of Plants, Place and Language*. Oxford: TrueHeart Press.

Sagan, Dorion, and Lynn Margulis. 1986. *Origins of Sex: Three Billion Years of Genetic Recombination*. New Haven, CT: Yale University Press.

Sale, Kirkpatrick. 2000. *Dwellers in the Land: The Bioregional Vision*. Athens: University of Georgia Press.

Salleh, Ariel Kay. 1984. "Deeper Than Deep Ecology: The Ecofeminist Connection." *Environmental Ethics* 6:339–45.

Sartore, Gina, Brian Kelly, Helen Stain, Glenn Albrecht, and Nick Higginbotham. 2008. "Control, Uncertainty, and Expectations for the Future: A Qualitative Study of the Impact of Drought on a Rural Australian Community." *Rural and Remote Health* 8:950.

Schweitzer, Albert. (1923) 1967. *Civilization and Ethics*. Translated by Charles Thomas Campion. London: Unwin Books.

Scofield, Bruce, and Lynn Margulis. 2012. "Psychological Discontent: Self and Science in Our Symbiotic Planet." In *Ecopsychology: Science, Totems and the Technological Species*, edited by Peter H. Kahn Jr. and Patricia H. Hasbach, 219–40. Cambridge, MA: MIT Press.

Scruton, Roger. 2012. *How to Think Seriously about the Planet: The Case for Environmental Conservation*. Oxford: Oxford University Press.

Seddon, George. 1997. *Landprints: Reflections on Place and Landscape*. Cambridge: Cambridge University Press.

Seed, John, Joanna Macy, Pat Fleming, and Arne Naess. 1988. *Thinking Like a Mountain: Towards a Council of All Beings*. Philadelphia: New Society Publishers.

Simard, Suzanne. 2016. "Notes from a Forest Scientist." In *The Hidden Life of Trees*, by Peter Wohlleben, 247–50. Vancouver: Greystone Books.

Smith, Daniel B. 2010. "Is There an Ecological Unconscious?" *New York Times Magazine*, January 27, 2010. https://www.nytimes.com/2010/01/31/magazine/31ecopsych-t.html.

Smith, Kirk R., and Alistair Woodward. 2014. "Human Health: Impacts, Adaptation, and Co-benefits." In *Climate Change 2014: Impacts, Adaptation, and Vulnerability*, 709–54. Cambridge: Cambridge University Press. https://www.ipcc.ch/pdf/assessment-report/ar5/wg2/WGIIAR5-Chap11_FINAL.pdf.

Sobel, David. 1996. *Beyond Ecophobia: Reclaiming the Heart in Nature Education*. Great Barrington, MA: Orion Society.

Sprinkle, Annie, and Elizabeth M. Stephens. 2011. "Ecosex Manifesto." SexEcology. http://sexecology.org/research-writing/ecosex-manifesto/.

Stanner, William Edward Hanley. 2009. *The Dreaming and Other Essays*. Melbourne: Black Inc. Agenda.

Steiner, Rudolph. (1923) 1998. *Bees: Lectures by Rudolph Steiner*. Translated by Thomas Braatz. Hudson, NY: Anthroposophic Press.

Steiner, Rudolph. 2008. *Spiritual Ecology: Reading the Book of Nature and Reconnecting with the World*. Translated by Matthew Barton. Forest Row, UK: Rudolf Steiner Press.

Sterling, Bruce. 2009. "Transcript of Reboot 11: Speech by Bruce Sterling." *Wired*, February 25, 2009. https://www.wired.com/2011/02/transcript-of-reboot-11-speech-by-bruce-sterling-25-6-2009/.

Sterling, Bruce. 2017. "A+E 2017—Bruce Sterling: Speculations." YouTube. November 7. https://www.youtube.com/watch?v=dfjHjJ2SFLI.

Stetka, Bret. 2016. "The Human Body's Complicated Relationship with Fungi." *Shots Health* News from NPR. April 16. https://www.npr.org/sections/health-shots/2016/04/16/474375734/the-human-body-s-complicated-relationship-with-fungus.

Taylor, Bron, ed. 2013. *Avatar and Nature Spirituality*. Waterloo: Wilfred Laurier University Press.

Thomashow, Mitchell. 1999. "Toward a Cosmopolitan Bioregionalism." In *Bioregionalism*, edited by Michael McGinnis, 121–32. London: Routledge.

Thompson, Clive. 2007. "How the Next Victim of Climate Change Will Be Our Minds." *Wired*, December 20, 2007. https://www.wired.com/2007/12/st-thompson-27/.

Tickell, Crispin. 2006. Foreword to *The Revenge of Gaia*, by James Lovelock. London: Penguin.

Toffler, Alvin. 1975. *The Eco-Spasm Report*. New York: Bantam Books.

Trappe, James M. 2005. "A.B. Frank and Mycorrhizae: The Challenge to Evolutionary and Ecologic Theory." *Mycorrhiza* 15 (4): 277–81. doi:10.1007/s00572-004-0330-5.

Tschakert, Petra, and Raymond Tutu. 2010. "Solastalgia: Environmentally Induced Distress and Migration among Africa's Poor Due to Climate Change." In *Environment, Forced Migration, and Social Vulnerability*, edited by Tamer Afifi and Jill Jäger, 57–69. New York: Springer.

Tuan, Yi-Fu. 1974. *Topophilia: A Study of Environmental Perception, Attitudes, and Values*, Englewood Cliffs, NJ: Prentice-Hall.

Tuck, Eve, Marcia McKenzie, and Kate McCoy. 2014. "Land Education: Indigenous, Post-colonial, and Decolonizing Perspectives on Place and Environmental Education Research." *Environmental Education Research* 20 (1): 1–23. doi:10.1080/13504622.2013.877708.

Tüür, Erkki-Sven. 2017. "Royal Concertgebouw Orchestra Performs World Premiere by Erkki-Sven Tüür." Royal Concertgebouw Orchestra Amsterdam. November 24. https://www.concertgebouworkest.nl/en/vincent-cortvrint-to-appear-as-soloist-in-brand-new-piccolo-concerto.

United Nations. 2016. *UN Harmony with Nature Report on Earth Jurisprudence.* http://www.un.org/en/ga/search/view_doc.asp?symbol=A/71/266.

Verplanken, Bas, and Deborah Roy. 2013. "'My Worries Are Rational, Climate Change Is Not': Habitual Ecological Worrying Is an Adaptive Response." *PLOS ONE* 8(9): e74708. https://doi.org/10.1371/journal.pone.0074708.

von Humboldt, Alexander. 1995. *Alexander von Humboldt: Personal Narrative.* Abridged and translated by Jason Wilson. London: Penguin Group.

Walker, Brian H., and David Salt. 2006. *Resilience Thinking: Sustaining Ecosystems and People in a Changing World.* Washington, DC: Island Press.

Wallace-Wells, David. 2017. "The Uninhabitable Earth: Famine, Economic Collapse, a Sun That Cooks Us: What Climate Change Could Wreak—Sooner Than You Think." *New York Magazine*, July 9, 2017. http://nymag.com/daily/intelligencer/2017/07/climate-change-earth-too-hot-for-humans.html.

Warsini, Sri, Kim Usher, Petra Buettner, Jane Mills, and Caryn West. 2015. "Psychosocial and Environmental Distress Resulting from a Volcanic Eruption: Study Protocol." *Collegian* 22:325–31.

Watts, Jonathan, and John Vidal. 2017. "Environmental Defenders Being Killed in Record Numbers Globally, New Research Reveals." *Guardian*, July 13, 2017. https://www.theguardian.com/environment/2017/jul/13/environmental-defenders-being-killed-in-record-numbers-globally-new-research-reveals.

Watts, Nick, W. Neil Adger, Paolo Agnolucci, Jason Blackstock, Peter Byass, Wenjia Cai et al. 2015. "Health and Climate Change: Policy Responses to Protect Public Health." *Lancet* 386 (10006): 1861–1914. http://www.thelancet.com/journals/lancet/article/PIIS0140-6736(15)60854-6/references.

Weisberg, Barry, ed. 1970. *Ecocide in Indochina: The Ecology of War.* San Francisco: Canfield Press.

Weldon, Annamaria. 2014. *The Lake's Apprentice.* Perth: UWA.

Westaway, Michael, Jon Olley, and Rainer Grun. 2017. "Aboriginal Australians Co-existed with the Megafauna for at Least 17,000 Years." *The Conversation*, January 11, 2017. https://theconversation.com/aboriginal-australians-co-existed-with-the-megafauna-for-at-least-17-000-years-70589

Wilson, Edward O. 1984. *Biophilia.* Cambridge, MA: Harvard University Press.

Wilson, Edward O. 1992. *The Diversity of Life.* London: Penguin.

Wilson, Edward O. 2016. *Half Earth: Our Planet's Fight for Life.* New York: Liveright.

World Commission on Environment and Development. 1987. *Our Common Future.* Oxford: Oxford University Press.

World Health Organization. 2017. *Mental Disorders Fact Sheet.* http://www.who.int/mediacentre/factsheets/fs396/en/.

Worthy, K. 2016. "Soliphilia and Other Ways of Loving a Planet: Can It Help to Name Our Love for Earth and Our Despair for Its Destruction?" *Psychology Today,*

May 9, 2016. https://www.psychologytoday.com/blog/the-green-mind/201605/soliphilia-and-other-ways-loving-planet.

Wright, Alexis. 2013. *The Swan Book*. Artarmon, New South Wales: Giramondo.

Wright, Christopher, and Daniel Nyberg. 2016. *Climate Change, Capitalism, and Corporations: Processes of Creative Self-Destruction*. Cambridge: Cambridge University Press.

Wright, Frank Lloyd. 1958. *The Living City*. New York: Meridian Book.

Yoder, Sarah. 2018. "Assessment of the Potential Health Impacts of Climate Change in Alaska." Alaska Division of Public Health. January 8. http://www.epi.alaska.gov/bulletins/docs/rr2018_01.pdf.

Zyga, Lisa. 2017. "Digitally Printed Cyanobacteria Can Power Small Electronic Devices." Phys.org. November 27. https://phys.org/news/2017-11-digitally-cyanobacteria-power-small-electronic.html#jCp.

Index